AF422085

CHRISTIANS THAT FOUGHT SLAVERY

Christians That Fought Slavery

Rebekah Brewster

Quietbeauty Publishing

Contents

Copyright © 2022 by Rebekah Brewster

All rights reserved. No part of this book may be reproduced in any manner whatsoever without written permission except in the case of brief quotations embodied in critical articles and reviews.

First Printing, 2022

This book is dedicated to YOU,
as you stand for the truth,
no matter what the world throws at you.

*"Be strong in the Lord and in His mighty power.
Put on the full armor of God
so that you can make your stand against the devil's schemes."*
Ephesians 6:10-11 (BSB)

<u>Also by Rebekah Brewster:</u>

Walk With God: 25 True Stories
Troublemakers: Volume 1

Legal Disclaimer

This book is history and entertainment. True stories of real people finding their way in life.

<u>This book is NOT legal, tax, investment or medical advice.</u>

Any description in this book of violence or bad behavior is so we can learn from history. Please don't use anything in this book to make a bad decision or commit violence. History needs to be known so we study it but we do not condone any of the wickedness studied in this book.

Throughout this book, each historical figure speaks for themselves, because we need to know our history.

This book is the author's personal opinion. The author does not speak for any of the people or companies mentioned in this book. This book was researched by the author and believed to be true. Please understand that even historians still disagree over what happened in history so there are no guarantees in this book.

Quotes in this book are either in the public domain, used with permission, or used under Fair Use of the U.S. Copyright Act. Some grammar rules were broken for emphasis in writing this book. Also, please note that since the English language has evolved considerably over the last several hundred years, some quotes have been edited to reflect modern English as well as for clarity and brevity. Now let's dive right in.

Image Credits

As long as evil existed in the world, good people were fighting it.

Once you read the story of Lafayette, you'll know why he is on both the front and back cover. During the darkest days of the Revolutionary War, when the military lacked warm clothing and shoes, Lafayette sacrificed everything to help.

The front cover image is John Ward Dunsmore's 1907 painting showing Lafayette with George Washington at Valley Forge during the Revolutionary War. They are looking at bloody footprints in the snow because many of the soldiers are having to walk without shoes.

The back cover is the 1909 print by Percy Morgan, *"Lafayette's Baptism by Fire,"* showing him leading troops into battle.

Lafayette was a devout Christian who fought slavery as you will read about soon.

All pictures in this book are courtesy of the Library of Congress unless otherwise stated. Images have been left in their original state to show the history as it was.

This book tells powerful stories of Christians who fought a horrible evil. Yet to understand our history we will also study some dark history of people who did evil.

We will study the horrible evil of George Washington being a slaveholder and the Christians who dared to confront him on his sin.

Note to Readers

For much of American history, the people who wanted to destroy slavery were called:
<u>Abolitionists</u>

1

Prologue

Years ago I was looking through some old books from the 1800's when something caught my eye.

History of the Rise and Fall of the Slave Power by Henry Wilson.

Henry Wilson (1812-1875), served 18 years in Congress as the Senator from Massachusetts. He was the driving force behind some of the earliest civil rights legislation in America. He was also a strong Christian who believed in standing for the truth.

He had written a book to tell future generations of the sacrifices that had been made for their freedom. He wrote about how during the Civil War, no one knew which side would win. People would nervously open their newspapers, dreading the latest news.

He wrote how the evil system of slavery was so powerful no one thought it would ever be destroyed. Then how amazing it was when the day finally came that the evil system was crushed beneath the feet of the American people, who did not back down until the principles of the *Declaration of Independence* became the law of the land. Reading his book inspired me to study the characters and I found some powerful stories that need to be told. I hope this book will inspire you to never forget how precious freedom is.

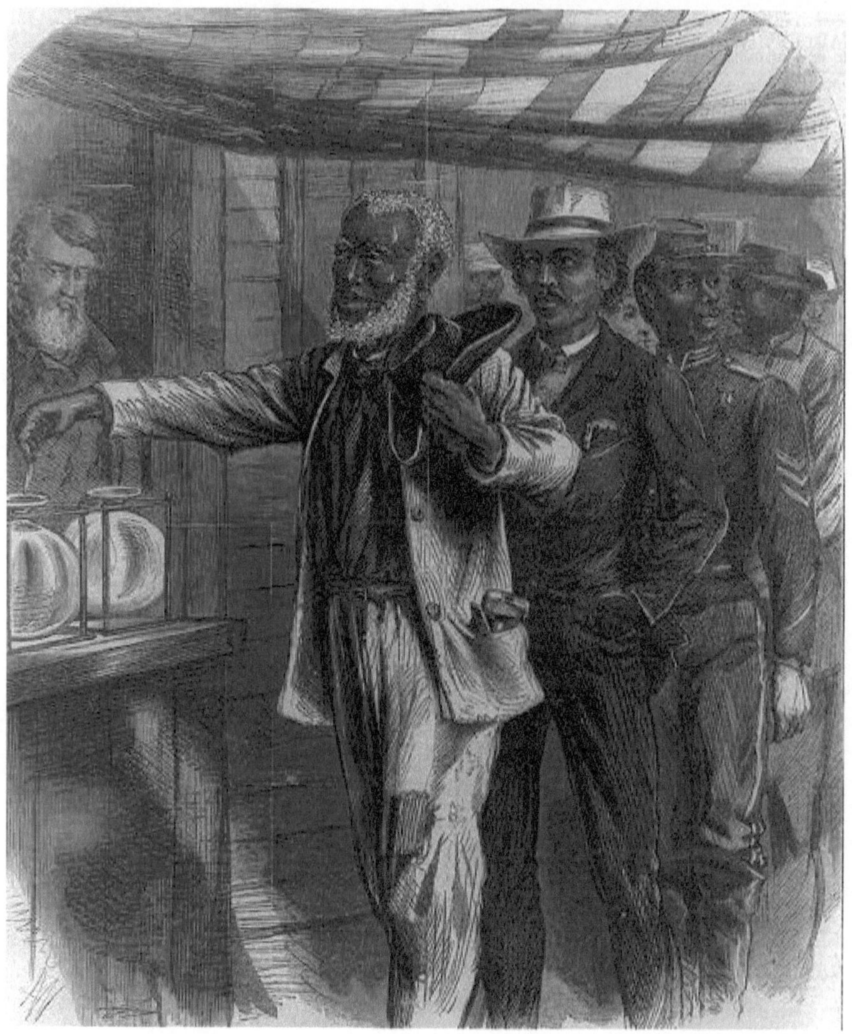

Drawing showing the Fifteenth Amendment in action. Artist
is unknown. First appeared Nov 16, 1867 in Harper's
Weekly.

Historical Note:

- Thirteenth Amendment to Constitution: Destroyed slavery.
- Fourteenth Amendment to Constitution: Gave equal protection under the law to everyone, especially former slaves.
- Fifthteenth Amendment to Constitution: Confirmed voting rights of former slaves.

2

William Wilberforce

"Until the time that his word came to pass,
The word of the LORD tested him."
Psalms 105:19 (NKJV)

William Wilberforce
1759-1833
From England

William Wilberforce was a very busy man. At 21 years old, he became one of the youngest people elected to Parliament. His time was consumed by official duties and state dinners. Yet he always had time for God. No matter where he went or what had to get done, he stayed focused on seeking God's heart.

He was a man of prayer, often taking long walks in his garden, praying for his nation. Yet when he finally heard from God about his future—it was a massive project.

On October 10, 1787, Wilberforce wrote in his journal:

"God Almighty has set before me two great objects—The suppression of the slave trade and the reformation of manners."[1]

Wilberforce's goal felt impossible. How could he take down the massive worldwide empire of evil? The institution of slavery had ruled the world for much of human history. While many had tried to resist and several slave revolts had succeeded, no one had ever been able to destroy this monster.

The more he thought about it, the more impossible it seemed. He would have to wait for the right opportunity. While he waited, he had a very comfortable life to enjoy.

Born into an affluent family, he had known nothing but luxury. His days were filled with parties in England's highest social circles. Though he had the life everyone would want, all he really wanted was to please God's heart.

While most people only dream of the wealth and power Wilberforce had, he felt very guilty about it. Wrestling with his conscience, he felt like the only way he could please God was by giving up everything to find some remote place to serve in the ministry.

His best friend didn't like his idea. Just like him, William Pitt had been born into the top of British society but Pitt understood that his destiny was to serve in politics. For many years he had prepared for this future, knowing someday he would be able to make a difference. In 1780, he ran for Parliament and lost. He just waited and prayed, knowing things would change. The opportunity would come sooner than he thought.

The Revolutionary War ripped apart political alliances in England. Long term friends became bitter enemies as some people supported America and others didn't. Throughout the political upheaval, Pitt stayed on the good side of the King. He was rewarded with being appointed as Cabinet Minister.

Pitt impressed the King with his strong people skills and sense of humor. Soon the King promoted him. At only 24 years old, Pitt became the youngest Prime Minister in England's history.

As Pitt rose to power, he stayed close to Wilberforce, often seeking advice from him.

Wilberforce described, *"For months at a time I spent every morning with him, while he was transacting business with his secretaries. He confided in me so much that he talked freely of people, plans, projects, and speculations."* [2]

In 1785, as Wilberforce experienced his crisis of faith, Pitt tried to change his mind. They were both young and in power. Working together in politics—couldn't they change England for the better?

Wilberforce disagreed. He insisted on walking away from his money and power. So Pitt tried one last time to change his mind by saying, *"You are deluding yourself into principles which will work against your goal, making you end up useless. Surely being a Christian leads to action not just meditation."* [3]

Wilberforce still wasn't convinced. He just wanted to find God's plan for his life. Seeking more advice, he decided to visit a close family friend. Wilberforce would have a life changing conversation with John Newton, the pastor who wrote the song *Amazing Grace*.

As Wilberforce was growing up, John Newton had been a friend and frequent visitor to the Wilberforce family home. Wilberforce had attended services at Newton's church. Many times he had heard the story of how God had done an amazing work in Newton's life.

As a young man, Newton was forced into the British Navy. When he rebelled, he was brutally punished. Still Newton disobeyed orders. Eventually the ship's captain had enough and decided to abandon him. When the ship docked at a remote port in Africa, Newton was kicked to the curb. There he found a job in the slave trade and began making a lot of money.

He never thought about God, until one day at sea, his ship ended up in a terrible storm. When the boat looked like it was going to sink, he cried out to God for help. The boat stayed afloat. They made it safely home and Newton started searching to know more about God.

Years passed. As Newton grew in his faith, the time came when he repented of the evil he had done and became an abolitionist.

After being formally ordained in 1764, he pastored a church in London. At that church, Wilberforce got to know him really well.

While Wilberforce had visited Newton for advice many times, the conversation they had in December 1785 changed Wilberforce forever.

Wilberforce told Newton, he wanted to be like him and dedicate his entire life to serving in the ministry. Wouldn't God be pleased if he gave up everything?

Newton listened carefully and then strongly disagreed. God had put Wilberforce in Parliament for a reason. He needed to focus on serving the Lord—where he already was—with what he already had.

Think about what he could do for God in Parliament. Very few people had access to the highest circles of power in the nation. Newton told him to think about how Joseph and Daniel in the Bible were called by God into high governmental positions.

Newton warned him against the idea of choosing a life of solitude in some far away place. The years he had spent building relationships were building towards a future that would touch many lives. Newton warned him not to walk away from his social life when the social life was a gift from God.

Newton told him, *"The Lord teaches those that seek Him, the direction they should walk. He doesn't reveal every part at once. Nor does he reveal it through the rules we make for ourselves. It is gradually revealed through experiences."*

"Stay focused on fellowshipping with God. Only give up what dampens your spirit from seeking Him. Let love guide you to the middle path between the extremes of unnecessary sacrifice and wrong obedience."[4]

Wilberforce went home greatly comforted. He wrote in his diary, *"My religious opinions have changed. Now I understand my duty. My walk is a public one. I must socialize with people or lose the work which God has assigned me."*[5]

God's plan for him was already in motion.

Wilberforce had a friend in Parliament, Sir Charles Middleton, who was an abolitionist.

Disturbed by how the other abolitionists in England were trying to work by themselves, Middleton tried to get them to work together.

Together with his wife Margaret, he started inviting abolitionists to their house for parties. Wilberforce was always invited. One of the people he met there was James Ramsey.

Ramsey was a highly educated doctor who had served in the British Navy under Middleton's command, way back when Middleton had once been Captain of the ship *Arundel*.

When Ramsey suffered a broken leg from a fall onboard ship, he retired from the Navy and entered the ministry. He moved to the British colonies to preach the Gospel. There he was shocked to see the horrible evil happening on the plantations.

Trying to destroy this evil, he lobbied the British government for social justice. His efforts were blocked by the rich plantation owners.

When he returned to England, Ramsey wrote a book describing what he had seen. His book shocked England. No one was supposed to talk about what actually was happening thousands of miles away in the British colonies.

BACKGROUND INFO

Spain, France, Holland and England had been establishing colonies around the world and competing for new territory.

In 1492, Spain sent Christopher Columbus to find a new trade route to the east.

When Columbus returned with descriptions of a new land, Spain sent several more explorers. One of those explorers was a black man named Juan Garrido (1480-1550).

In 1503, Garrido went to the island of Hispanola (now known as Haiti and the Dominican Republic). Then he joined Ponce De Leon in exploring Florida, Puerto Rico, and Cuba. Later he explored California. Then he settled in Mexico, becoming the first person to plant wheat in the new world.[6] His farm used slave labor. Garrido also helped Hernando Cortez conquer the Aztec Empire.

When Cortez landed in Mexico, he was horrified to find hundreds of thousands of human bones from people sacrificed to appease the gods of the Aztec Empire. The smell was terrible as basins were full of human blood. Buildings had been built with bricks made from human bones.

The people sacrificed were often captives, which the Aztecs had captured while fighting local tribes. This had turned the local tribes against the Aztecs, enabling Cortez to recruit them to help overthrow the Aztec empire and end the bloody rituals.

During this time period, there was a black man who explored America. His name was Luis (last name unknown). He worked as a sailor on a merchant ship. The weather was rough, causing him to end up shipwrecked on the coast of Florida around 1554. He moved in with the Native Americans, learning their language and absorbing their culture.

When Spanish explorer Pedro Aviles came to Florida, in 1565, Luis' skills as a translator helped Aviles negotiate with the natives.[7]

Other black men also came to America as explorers and conquerers for Spain, going with Hernando de Alarcon and Francisco Vasquez in travels through Arizona and New Mexico in the 1540's.

As Spain explored new territories, they opened gold and silver mines and forced the natives to work them.

When the native population began to die off from the harshness of slavery, Spain replaced them by importing African slaves to work in its colonies. Taking control of a vast amount of new territory including Mexico, Florida, Puerto Rico, Cuba, Venezuela, Guatemala, Argentina and Peru, Spain made a fortune.

Watching Spain become rich, the rest of Europe tried to catch up in the race for new territory.

In 1541, France sent four hundred settlers to Canada. They did not last long as the harsh winter made the people give up and go back home. French settlers continued exploring Canada, making a permanent colony in 1605 at Nova Scotia. Setters soon spread from Canada down the Mississippi River into Louisiana Territory. To develop this land, France also brought slavery to America, opening trading posts in Africa to ensure enough supply.

Holland wasn't too far behind.

In 1615, Holland launched a permanent colony on the Hudson River. They continued expanding by purchasing the island of Manhattan. By 1637, they were importing slaves from Africa.

The Dutch negotiated directly with the leaders of Africa to establish one of the largest slave trading outposts. By 1642, they had exchanged ambassadors with the Kingdom of Congo. They also signed trade contracts with King Garcia II of Congo. As diplomatic relations were established, one of the first ambassadors from Congo to Holland was Miguel de Castro.[8]

During this time, there is an interesting story.

In 1648, Massachusetts Gov. John Winthrop wrote in his journal about a free black man from Holland (name unknown) who came to America to work for the Governor of the Dutch colony in New York. While living there, he became a missionary to the Indians.

Winthrop wrote, *"Having learned some of their language, he began to teach them the things of God and the Lord blessed his endeavors. The Indians listened to him and began attending his regular Bible studies."*[9]

No other details are known about him.

In 1562, England joined the Atlantic slave trade when Sir John Hawkins began his voyages. He took slaves from Africa to Haiti, traded them for sugar, then brought the sugar back to England to sell for enormous profits. That ten month voyage made him a lot of money.

To his surprise, he found Queen Elizabeth was not happy about it. She summoned him, protesting that this vile trade *"Would call down the vengeance of Heaven."*[10]

Hawkins still continued in the evil trade until his third voyage, *"Terminated most miserably."*[11]

Driven off course by a hurricane, he was attacked by Spanish ships in the Gulf of Mexico. Many of his crew died in battle. Some were captured and executed. Others were sold into slavery. A few escaped, including Hawkins himself. He continued exploring the new world, until dying of disease in Puerto Rico.

(Wilberforce later showed evidence in a speech to Parliament that slave traders had a much higher death rate than regular sailors working on ships due to disease and getting attacked by enemies.)

The slave trade continued without him as British merchants made money, transporting slaves to Spain's rapidly growing colonies.

England developed its own colonial empire across islands in the Caribbean including Jamaica, Barbados and the Bahamas.

In 1713, Spain and England joined together to protect their profits. They signed a treaty, giving British ships the monopoly on transporting slaves to the colonies of both Spain and England.

Meanwhile, there were many people trying to destroy this evil trade. Christian abolitionists in England, relentlessly lobbied Parliament to abolish the trade.

When Parliament wouldn't listen, the abolitionists filed lawsuits and fought through the court system. They proved that slavery was illegal under British law dating all the way back to the *Magna Carta.*

In 1772, they won a major victory when a British court ruling declared, *"As soon as any slave set his foot upon English territory, he became free."*[12]

This was the famous Somerset case. James Somerset was kidnapped in Africa, sold in America and taken by his master to England. There he ran away and approached Granville Sharp for help. Sharp filed a freedom lawsuit, resulting in the landmark court decision declaring that British law had never actually legalized slavery.

Granville Sharp led the abolitionist movement, writing books showing from the Bible how much God hated slavery. His books were read around the world, even finding their way into the personal library of Founding Father John Adams.[13]

When Adams traveled to England to serve as the U.S. Ambassador, he got in contact with Sharp. Adams asked Sharp for a favor: to send his abolition books to a young French nobleman named Lafayette.

While Sharp's book changed many hearts and minds, the abolitionists could not get past the political system standing in the way. The court ruling Granville had won—would only be able to free the few slaves who had moved with their owners to England itself.

It would not be able to reach the thousands of slaves suffering in the British held territories. Too many members of Parliament were getting rich off those colonies to be willing to interfere with them.

By the time that William Wilberforce was born, the slave trade was controlled by the very deep pockets of the rich and powerful.

Yet no matter how impossible it seemed, there were always people fighting the battle to bring freedom.

In 1776, David Hartley introduced the first motion in Parliament that *"The slave trade was contrary to the laws of God and men."*[14]

He was bitterly opposed. People thought he was a fool for trying to oppose one of the most lucrative industries in the world.

To keep him too busy to interfere, he was assigned to supervise negotiations with America as the Revolutionary War began.

When the war ended, he negotiated a peace treaty with diplomats from America: John Adams, Benjamin Franklin, John Jay and Henry Laurens.

As they discussed the terms of the agreement, Hartley realized that he was dealing with Americans who shared his point of view.

John Jay suggested adding antislavery language to the treaty such as: *"British subjects be forbidden to import slaves to America from any part of the world, it being the intention of the United States to entirely prohibit the importation thereof."*[15]

Harley loved that idea. Yet England refused to allow America to put any limitations on its shipping industry. That clause was struck out of the treaty, but the men who had negotiated it returned home and continued to fight slavery in a different way.

John Adams had refused to own a slave in a time when neighbors thought he was crazy. How could he pay people to work for him, when labor could be had for free? No matter how many people didn't like him, Adams refused to sacrifice his principles. In his early years as a young attorney, in 1769, he helped a slave win a freedom lawsuit in Massachusetts. Years later, he wrote the Massachusetts Constitution in a way that enabled the state courts to abolish slavery in court decisions just after the Revolutionary War. By the time that the first U.S. Census was taken in 1790, it recorded there were no slaves left in Massachusetts.

The second member of the team, Benjamin Franklin, freed the slaves he owned. He was also elected as President of the first antislavery group in America. Just a few days before he died, his last public act was to petition Congress to destroy the trade.

The third member of the team, John Jay, went home to New York and introduced legislation in the State Legislature to abolish slavery.

Jay wrote, *"Men held as slaves by the laws of New York are free by the law of God."*[16]

When his legislation was voted down, he formed a powerful abolitionist group. They led the battle until New York formally abolished slavery in 1799. By then, Jay had become the Governor of New York, who signed the historic bill.

The last member of that negotiating team, Henry Laurens, was a ruthless slave trader. He never understood why it was wrong. That didn't stop his son, John Laurens, from fighting for freedom.

During the Revolutionary War, when America's military needed soldiers, John Laurens lobbied George Washington to recruit slaves for the military. His reasoning was *"To advance those who are unjustly deprived of their human rights and to open the door for emancipation."*[17]

Since each colony was responsible to produce soldiers, Washington sent John to the legislature of his home state of South Carolina to persuade them to try his idea.

They flatly refused. When Washington heard he wrote to John, *"I'm not surprised that your plan failed. That spirit of freedom, which was willing to sacrifice anything at the beginning of this war, has long since subsided. Now selfishness is driving people."*[18]

During the Revolutionary War, John Laurens sacrificed his life in battle at age twenty-seven. Only then his father honored his wishes to free the slaves that he would have inherited.

While this was happening in America, back in England, William Wilberforce was a young man pursuing God's will for his life.

WILBERFORCE STARTS HIS JOURNEY

When Wilberforce met Ramsey and heard about Ramsey's travels, he was disturbed to find out the great evil happening thousands of miles away in the British colonies.

When Ramsey published his book, Wilberforce read it. In a time when those that knew, had been threatened into silence, Ramsey had just blown the lid off the evil system.

Years of carefully woven propaganda designed to hide the truth had been obliterated by his eyewitness testimony. Ramsey's book was read throughout England, waking the people up. Many people were deeply moved by his book, including powerful people like Wilberforce and Pitt.

One morning, while Wilberforce and Ramsey were eating breakfast at the Middleton house, Margaret Middleton turned to her husband Charles. She told him to bring an antislavery bill in Parliament.

Charles refused. How could he debate such an explosive topic when he was terrible at public speaking? Feeling inadequate against the heavy opposition he would face, Charles said someone else should do it.

Margaret turned to Wilberforce and asked him to introduce it.

Wilberforce took some time to think about it, as her suggestion became *"One of many impulses which were all giving my mind the same direction."*[19]

How could Parliament pass a law against slavery when there were so many roadblocks in the way? Any new laws had to be passed by a majority of both houses of Parliament.

The House of Lords, which had inherited their seats from their wealthy families, was more powerful than the House of Commons which had been elected by the people.

The House of Lords could overrule anything decided by the House of Commons. Even if both Houses of Parliament could be persuaded to pass antislavery legislation, the King could overrule any new law at any time. Power was in the hands of a few people, many of them were directly benefiting from the trade.

Wilberforce decided that before he opened his mouth, he needed to know more about it. Launching a full investigation, he carefully studied the situation from every angle.

He approached merchants working in the slave trade and asked all sorts of questions. Because he was a powerful man in Parliament, they sat down and talked with him.

He wrote, *"I found the merchants willing to freely give me information, because at that time their trade was not alarming the nation. But their descriptions were full of errors and bias."*[20]

He also researched any reports or other documentation he could find. Working together with a team of researchers, he spent months studying reports, eye-witness accounts and documentation.

Friends who came to visit him, began to worry about his health, noticing that he never stopped working on it from morning until night. The more he learned, the more he hated *"The enormous, dreadful wickedness."*

Wilberforce decided, *"No matter what the consequences, I would never stop until I had destroyed it."*[21]

After months of collecting evidence, in May 1787, he went to stay at Pitt's home for a few days, so they could strategize together. Pitt owned a beautiful home on a large estate with a meadow, where they could relax and talk in private.

Together they walked through the meadow and sat down under an oak tree, talking about all the different ways they could do this.

Wilberforce didn't feel ready yet.

Pitt replied, *"Why don't you introduce a motion in Parliament to examine the slave trade? You've already taken great pains to collect the evidence."*[22]

Introducing it was difficult. Getting it passed felt impossible.

Knowing he needed the support of as many Parliament members as possible, Wilberforce took several more weeks, approaching his friends to ask for their vote.

Working with as many politicians as possible, he gathered support for his bill. He also tried to stir up public opinion on the subject. Helping him was a newly launched British antislavery group.

Finally the abolitionists were working together. Several community leaders had began meeting regularly to strategize how to destroy slavery. The chairman was Granville Sharpe, who had won the Somerset court case. He planned strategies for attacking every aspect of the evil trade. He was thrilled to be working with Wilberforce. *His position, influence, and reputation are the biggest advantage to our cause.*[23]

(These meetings birthed Anti-Slavery International, which still exists today. In 2008, they won a major court decision, holding the government of Niger responsible for failing to protect a girl named Hadijatou Mani, from being enslaved. They are still working around the world to eradicate all forms of slavery and have successfully freed thousands of oppressed workers in places like Nepal, Brazil and Mauritania.)

Wilberforce asked this group to focus on a public relations campaign to make Parliament realize that the British people wanted to destroy slavery. The group began gathering signatures on antislavery petitions to present to Parliament. Within months, they had found over 100,000 people to sign petitions.

Wilberforce presented these petitions to Parliament, demanding something be done.

Parliament listened. These petitions represented a broad spectrum of every level of British society. One from the University of Cambridge declared, *"Firm belief in a kind Creator assures us that no system founded on the oppression of one part of mankind could be beneficial to another."*

While public pressure was mounting on Parliament, Wilberforce became seriously ill.

All the endless hours of reading report after report had taken its toll on his health, forcing him to take time off to seek medical treatment. Not knowing when Wilberforce would recover, Pill volunteered to take the lead.

On May 9, 1788, Pitt ordered the Privy Council of Parliament to conduct a full investigation on the slave trade. Parliament adjourned before getting it done.

By the time they reconvened, Wilberforce had fully recovered and was back in his seat.

Almost a year later, in April 1789, the Privy Council presented its report. It provided many statistics, such as the high death rate among the sailors. More British sailors had died working in the trade than on any other merchant ships. However, the report also had misleading testimony from slave traders who painted a rosy picture of deceptive propaganda, trying to justify their evil deeds.

Listening to the report, Wilberforce took notes on how he would challenge it. Then on May 12, 1789, he presented the results of his own investigation. Exposing their lies, he completely shredded the report.

Members of Parliament listened in shocked silence as he thundered, *"The wickedness is so enormously dreadful that we can stop at nothing but complete abolition. Slave owners claim that they treat their slaves well, because that's in their best interest. That's ridiculous! Why do we make laws? Because people cannot be trusted to act against their own selfish interests!"*

Then turning towards the religious Parliament members who had tried to twist the Bible to justify the evil, he declared, *"God Almighty has forbidden the practice of shedding innocent blood. How could God possibly condone a trade founded on sin?"*

For over three hours, Wilberforce spoke with humility and strength. *"I'm not there to blame others. We are all guilty for having allowed this horrid trade to happen. We can't plead ignorance. We can't avoid it. Don't let Parliament become the only people insensitive to justice."*

When he had finished, Wilberforce sat down.

Then the debate began. Every type of excuse was made.

Parliament Member Mr. Penrhyn of Liverpool predicted that total economic ruin would from come from abolition. If it passed, all the mortgages on plantations in the West Indies would go into default, causing a massive economic meltdown.

Next Mr. Gascoyne of Liverpool stood up and accused Wilberforce of lying. He promised to disprove all of Wilberforce's facts. This was an empty promise as he had no evidence to present.

Abolitionist member Charles Fox couldn't wait to take the floor and confront Gascoyne. Fox roared, *"This evil trade has no place in a Christian country. If we don't abolish this, we are guilty of all the wickedness it produces!*

Mr. Alderman answered. *"While you are giving way to the goodness of your heart, be careful not to ruin your country. Abolition will bankrupt London. Let's just pass some regulations to make the trade more wholesome."*

Mr. Fox sneered, *"You can't regulate robbery and murder!"*

Alderman replied there just wasn't enough information to make a decision.

Turning to Wilberforce, Alderman asked, *"Do you have any more evidence to present?"*

Wilberforce shook his head. Alderman concluded, *"Then let's delay this debate until Parliament's next session."*

Pitt stood up and angrily complained about the *"Unnecessary delay of such an important issue."* Yet it would still have to wait.

When the next session came, Wilberforce was more than ready. He ripped through the lies and evil propaganda to show the truth.

Presenting fact after fact, he proved that abolition would benefit the economy. Free labor was much more productive for the economy than slavery. *"Not just morality and religion requires you to abolish this, but also sound economic policy. Let me argue this from the evidence itself."*

(Years later on July 12, 1858, during a House of Commons debate, it would be noted that England's international trade was thriving and had increased 87% after slavery was abolished.)

Wilberforce showed how a few years before, experts had predicted the complete collapse of the British economy, if America won its war for independence. Yet these fears were unjustified. England had continued to prosper despite losing one territory.

Then Wilberforce exposed the dark side of what was happening, presenting overwhelming evidence of *"Oppression, fraud, treachery, and bloodshed."*

He noted, *"In Africa, the natives never go anywhere without carrying a firearm. When Captain Wilson asked them why, they pointed to a slave ship sitting in the harbor."*[24]

As Wilberforce closed his speech, he implored them to *"Think about eternity and the future consequences of our choices today. What is there in this life that should make any man disobey the laws of God?"*[25]

"We will never quit until we have destroyed every trace of this bloody traffic which one day our future generations will look back at and wonder how it could have even existed."[26]

Colonel Tarleton disagreed. When Wilberforce finished, Tarleton stood up to give a long speech on the history of the trade noting how *"Parliament has always supported it."*

Arguing that *"Abolitionists are nothing but useless fanatics,"* Tarleton insisted, *"Abolition will ruin the economy, raise taxes, and increase the national debt."*

By attacking a *"Trade which employs 5,500 sailors and 160 ships, Wilberforce had better have a plan to save our economy."*[27]

Mr. Fox was furious. *"I can't believe you are so hard hearted as to vote for this vile trade and then go home to your families!!"*

Mr. Grosvenor fired back. *"Abolition is impossible. If we stop it, foreigners will continue the trade without using our humane regulations."*

James Martin replied, *"I had no idea how much selfishness could pervert your minds until we started this debate."*

"How can you deceive yourselves into believing you have a right to oppress anyone, when the rule of our religion is to do unto others as you would want to be done unto you?"[28]

Mr. Thurlow wasn't so sure. *"The French encourage the slave trade by offering subsidies. We ought to do the same. Let this subject rest until the next session."*

"Already one of my constituents came to me worried, saying, 'I am ruined if you pass this bill! I have risked everything I have, £30,000, on this trade.'"[29]

The debate dragged on all night.

Pitt gave a long speech reminding everyone that hundreds of years ago, when the Roman Empire ruled the world, it had also conquered England, forcing the British people into slavery. Now hundreds of years later, how could England persist in the very evil that their ancestors had been freed from?

Pitt: *"This is not a question of feelings, but justice. Do we not have laws against injustice? Here in England a pickpocket is condemned to death but we allow the much worse crimes of the slave trade to go unpunished."*[30]

After Pitt was finished, Mr. Fox also reminded them, *"Wilberforce has proven there is no danger in abolition. If we don't end this trade, we have forever ruined our legacy. There is only one way we can judge this, by the rule of Christianity. All mankind is equal in the sight of their Creator."*[31]

Then he described how wherever Christianity had spread it had pushed back slavery, until it was slowly destroyed.[32]

The debate continued all night. Finally, early in the morning, at 3:30AM the vote was taken. Wilberforce's motion lost by 163 to 88.

While Wilberforce had lost the first battle, he had succeeded in gaining the attention of the nation.

The team of abolitionists working with him continued the battle. Since one of the main products coming from the British plantations was sugar, they organized a sugar boycott.

Hundreds of thousands of British people stopped buying sugar to show their support for abolition.

Another half million people signed new antislavery petitions that were mailed to Parliament.

Wilberforce told his friends, *"We must rely on the feeling of the nation rather than the conscience of the House of Commons."*[33]

Wilberforce continued introducing abolition bills each session, knowing they would be rejected, but hoping that it would keep the nation focused on this issue. As the work continued, Wilberforce received an encouraging letter from John Newton.

"You are representing the Lord in a place where many know him not."[34]

As soon as he could, Wilberforce introduced another bill declaring *"It is the opinion of Parliament that this trade ought to be abolished."*

His bill was quickly voted down.

Wilberforce continued introducing abolition bills.

Parliament rejected them. Then finally Parliament passed one that said, *"The slave trade ought to be <u>gradually</u> abolished."*

Wilberforce was furious. How could they have changed his motion for immediate freedom into something that could be permanently delayed?

While people were congratulating him on what sounded like a victory, he wrote to a friend that he felt *"hurt and humiliated"* at how Parliament had passed *"A license to rob and murder."*[35]

Meanwhile, the plantation owners fought back. Meeting together to plan how to defeat Wilberforce, the proslavery opposition began working night and day to block his efforts. First they tried to destroy his reputation by accusing him of everything they could imagine. Wilberforce was accused of being a wife beater, when he didn't even have a girlfriend. Then they started spreading rumors he was having an affair with his housekeeper. They even falsely accused him of previously owning slaves in secret. The news media portrayed Wilberforce as someone who didn't care about the people of England.

Wilberforce ignored the personal attacks, burying himself in his work.

He continued introducing motions, but Parliament decided to leave the entire matter in the hands of the plantation owners themselves.

During this time, Wilberforce had an ally in Parliament named Daniel O'Connell. As a representative from Ireland, O'Connell had fought for Ireland to have greater authority over its own affairs. Whenever any bill came into Parliament concerning Ireland he was always the first on his feet.

One day he was visited at home by several Parliament members.

They said, *"We know that you've been working with Wilberforce. But if you'll shut your mouth on abolition, then whenever any bills come up in Parliament for Ireland you'll be sure of our fifty votes on your side."*

While the offer was tempting, O'Connell was insulted that they would try to buy his vote.

Staring them down, he thundered back, *"No! Let God care for Ireland! I'll never shut my mouth on abolition to save her."*[36]

While this was happening in England, it was about to be disrupted by the French Revolution.

For hundreds of years the people of France had suffered under the brutal system of feudalism. After severe famine lead to food shortages, the French people rose up. Out of the chaos rose a new ruler of France named Napoleon. He took control with an iron fist, ending the French Revolution. While he conquered much of Europe, he was not able to stop a slave revolt on the island of Haiti.

Thousands of miles away, the slaves in Haiti took their freedom in bloody warfare, led by General Toussaint L'Ouverture, a former slave with military experience. The revolt swept across Haiti and the Dominican Republic.

Thousands of people died. Napoleon tried to subdue the revolt, but the 25,000 soldiers he sent were easily defeated by the freed slaves. Haiti declared itself an independent nation and took its place among the nations of the world.

The world was watching as Napoleon gave up on Haiti, only to continue with his plan to conquer the rest of the world. Invading the nations closest to him, soon Napoleon controlled much of Europe.

The day the news reached England, that France had declared war on England, Wilberforce had actually been standing up in Parliament, ready to introduce another abolition bill.

Pitt sent a messenger, urging Wilberforce to sit down and wait while the country mobilized for war. Wilberforce was very frustrated at waiting on such an important issue, but knew there was no chance of passing the bill.

As he wondered what to do next, he received another letter from Pastor John Newton.

"The Lord has raised you up for the good of His church and the nation. While the position where God has assigned you is difficult, I know that He who has called you to it can strengthen you according to your day and time."[37]

Even though it was impossible to pass the bill, Wilberforce did not wait long before introducing it again. While the rest of the world was focused on dealing with Napoleon, Wilberforce kept fighting for abolition.

Every time his motions were voted down, he knew he was at least keeping the subject in front of the public. He had his Parliament speeches published and widely distributed, giving him a platform to let people know about the horrible suffering thousands of miles away in the British colonies. Just in case anyone missed his speeches, he also published a book on the slave trade, so the evil could not be hidden any longer.

Then Wilberforce got to work on the other assignment from God: to transform society's manners.

Wilberforce wrote the book *Real Christianity;* a powerful devotional that explored the difference between empty religious rituals and what actually pleased God's heart.

When the book was finished, Wilberforce tried to find a publisher.

No one wanted it. Publishers told him that books on religion didn't sell very well. The publishers refused to cover any of the costs, telling Wilberforce to pay for it himself. He did. Starting with a print run of 500 copies, Wilberforce self published his book.

To everyone's surprise, those 500 copies sold out in record time. Thousands more copies were printed. Those copies sold out so fast that bookstores could hardly keep it on their shelves.

His book was translated into several other languages and distributed around the world. Today his book is still in print and encouraging many Christians to go deeper with God.

Wilberforce wrote the book from his heart, apologizing for not being a better writer.

Emphasizing that real Christianity could change a nation, if people would focus on following Jesus instead of religious tradition he wrote: *"It cannot be denied that wherever Christianity has prevailed it has raised the general standard of morals."*

Encouraging everyone to read the Bible and get to know God for themselves, instead of just assuming that they were a Christian because they lived in a Christian nation, he wrote:

"Faith comes by hearing and hearing by the word of God. Yet even with the Bible in our houses, we are ignorant of its contents. How can we expect to be Christians without studying it? One day when we stand before God to give account of our lives, how will we be able to explain our willful ignorance?"

"Christianity calls us to practice the principles of Christ." Yet he worried that many Christians had lost the *"fear of grieving the Holy Spirit."*

Wilberforce believed firmly in the Bible. *"From the decision of the Word of God there can be no appeal."*

"We should cease to be deceived and confound the gospel of Christ with the systems of philosophers."

Then he grieved how many Christians had drifted from their faith. *"Their standard of right and wrong is not the standard of the gospel. They approve and condemn by a different rule."*

"They advance principles and maintain opinions opposite to the character of Christianity. Their opinions are not from the Word of God. The Bible lies on the shelf unopened and they would be ignorant of its contents except for what they hear occasionally at church."

He was also concerned about the decline of faith in England. *"In schools and colleges, Christianity is neglected."*

People were losing interest in going to church because of what had happened to the pastors. Church authorities were hiring friends and family to pastor. No experience or training was required. Thus, many pastors were drawing salaries without bothering to put in much effort at the church. Some rarely even showed up to preach. Others preached sermons that they had paid someone else to write. Some pastors collected salaries from two or three congregations while rarely appearing at any of them.

The spiritual needs of congregations were being neglected while no one held the clergy accountable. Wilberforce grieved how preachers talked about God while *"their lives are an open scandal."*

He urged them to think about eternity and how one day everyone *"shall stand before God to give account of all things done."*

He wrote about how some people used the excuse of human depravity to commit evil. *"They trifle with God's patience and despise His invitations."*

"St. Paul said it best, 'Because they did not like to retain God in their knowledge, he gave them over to a reprobate mind (Romans 1:28).'"

Yet while humans are born with a sin nature, *"This natural depravity shall never be an excuse for sin."*

He pointed to James 1:13. *"Let no man say when he is tempted, I am tempted of God."*

Wilberforce wanted everyone to know Hebrews 12:14: *"Without holiness no man shall see God."*

That holiness was *"hungering and thirsting after righteousness and to purify yourself even as God is pure."*

Wilberforce pondered Psalms 9:17. *"The wicked shall be turned into hell and all the nations that forget God."*

"There are wicked men—enemies of God—who draw others to commit evil."

He reminded them of what Jesus said in John 16:2, *"The time comes when whoever kills you thinks they do God service."*

"The Word of God teaches us not only to fight with our own depravity but also with the power of darkness and the evil spirit in the hearts of the wicked."

"Sensual gratifications have debased our nobler powers and blocked our hearts from willing obedience to God."

"Vice will finally terminate in misery."

He wrote how Christians need to be strong.

"Let them boldly assert the cause of Christ in an age when so many Christians are ashamed of Him."

"Let them pray continually for their country. Who can say but the Governor of the universe, who hears the prayers of his servants, may answer and avert our ruin."

"Fear not, though the world, the flesh and the devil are set in array against you. God is faithful."

"Beat the world at its own best weapons. Let your love be more affectionate, your mildness less open to irritation, your diligence more laborious, your activity more persevering."

"When all around him is dark and stormy, the Christian looks up at Heaven with hope and gratitude. Then no danger can alarm, no opposition can move, no provocations can irritate. The Christian's hope is not founded on the speculations or strength of man but on the Word of God who cannot lie. Believe in the Lord Jesus Christ and you shall be saved."

Wilberforce also dealt with prominent social issues of that time such as dueling. Young gentleman considered it their duty to avenge any perceived insult with a gentleman's duel.

These duels often resulted in one man killing another. While this brought much heartbreak to families, dueling was growing in popularity.

Wilberforce protested the absurdity of *"rushing into the presence of our Maker in the very act of offending Him,"* because *"pride is the chief cause of strife."*

Pointing out that dueling was *"deliberate preference of man's favor before the favor of God in an instance where our own life and another life is at stake,"* we should be more concerned with pleasing God's heart than protecting our reputation. *"As the Christian grows in grace, he grows in humility."*

"Seek the honor that comes from God. You cannot advance until you become indifferent to the praises of men."

He wrote about his own journey in learning not to give up the life that God had given him. How Christians should not be *"unfit for any social interaction and confined within gloomy walls."*

Explaining how important it was for Christians to go outside the four walls of the church and be involved in every level of society, he explained, *"True Christians desire to please God in all their thoughts, words and actions."*

"True Christianity is to devote the heart and life to God and a desire to know and fulfill His will."

"This is the love of God that we keep His commandments. This is the best standard to estimate the strength of religious affections."[38]

Throughout the whole book, Wilberforce emphasized that we are all responsible for our actions. We cannot sit around waiting for God to do something when He has given us the responsibility of seeking His will and making it happen.

Wilberforce's book would have a profound influence on many people, motivating them to take their walk with God seriously and begin seeking His will in their lives.

In addition to writing the book, Wilberforce also got involved with dozens of charitable organizations.

He created the first ever *Society for the Prevention of Cruelty to Animals.* The organization he launched is still around today, saving thousands of animals each year.

He also worked to overturn the harsh criminal laws in England, which had severely punished poor people for minor infractions. He fought to improve education in poor neighborhoods, even personally funding schools when no one else would. Becoming known as one of the most generous men in England, he poured his fortune into helping others.

Along the way he also fell in love. Barbara Ann Spooner, had been secretly admiring him from a distance. Everyone knew that she was of royal blood, being a direct descendant of King Richard III. However, she wasn't looking to marry in high society. She wanted someone who loved God.

Making the first move, she wrote Wilberforce a nice letter asking for his opinion on spiritual things and expressing her admiration for him. When Wilberforce received the letter, he was impressed by her boldness in approaching him.

They met in person, when mutual friends invited them both to a dinner party. It was love at first sight. Eight days later, they married on May 30, 1797. Wilberforce was thrilled to find his soul mate. Together they started a family, raising four boys and two girls.

In 1806, William Pitt became ill and passed away at the young age of 46. England grieved the loss of a great statesman.

Wilberforce grieved the loss of a dear friend who had been with him through thick and thin.

There was still a lot of work to do. Wilberforce continued to fight for abolition, but this time politics were aligning in his favor.

THE ANSWER COMES

The new Prime Minister of England was Lord Grenville: a close friend to Wilberforce. Grenville had been with Wilberforce in the meadow at Pitt's house when Pitt told Wilberforce to introduce the first abolition motion. Now he was in power and ready to open the way for the bill to pass. This time, they would try a completely different strategy.

Wilberforce was approached by another Parliament Member, James Stephen, who had developed the legal theory that just might be able to pass the bill.

His book *The War in Disguise* was already making a powerful impact in England. Stephen had realized that Napoleon was financing his dream for world domination by taking advantage of a loophole in the shipping industry.

Napoleon's wars cost a lot of money to fight. He was getting this money from the lucrative profits of the French colonial empire. Goods from these colonies were sold around the world and the profits shipped back to France.

After Napoleon declared war on Europe and began invading other countries, England had tried to disrupt Napoleon's supply lines with a naval blockade, to stop ships carrying cargo from going to and from the French colonies.

However, the naval blockade depended on the assumption that ships would only fly the flag of the country they represented. Napoleon had easily defeated the naval blockade by having all his ships fly American, British or other flags so that British ships would not attack them.

If England could pass a bill, making it illegal for British ships to transport anything to the colonies of a hostile nation, such as France, then they would have the right to stop and search any ship flying a British flag.

This would disrupt trade, cutting off Napoleon's funding. Once France could no longer access the resources of its colonies, Napoleon would run out of money, bringing an end to the war.

Stephen told Wilberforce that this was a subtle way of striking a huge blow at the slave trade by cutting off the Atlantic trade route being used to transport slaves from Africa to the European colonies. The plan was for Prime Minister Grenville to introduce the bill in the House of Lords where it might pass as a war measure.

Wilberforce was excited. This was the first glimmer of hope he had seen in a long time. Yet if proslavery Parliament members saw Wilberforce pushing a bill, then they would know it had something to do with abolition. Stephen told him to get his friend, Prime Minister Grenville, to support the bill and then stay out of the way.

Wilberforce talked to Grenville about the bill. Then he kept his mouth shut, as the bill began moving through the legislative process.

On May 7, 1806, Grenville introduced the bill into the House of Lords. *"If its obvious that we should not give advantages to our enemies, it was surely equally clear that we should not supply their colonies with slaves, thereby affording them means of cultivation contributing to increase the produce of their islands and thus enabling them to meet us in the market upon equal competition or perhaps to undersell us."[39]*

His arguments persuaded the House of Lords.

The bill passed by 43 to 18.

The bill moved to the House of Commons, where it was introduced by Charles Fox. After Fox gave a long speech on how this was one of the greatest moments of his career, Colonel Tarleton jumped to his feet to protest that this bill was nothing more than a trick by the abolitionists.

Tarleton stormed against it with a long speech trying to justify slavery.

Listening to the speech, Wilberforce sat in his seat and tried to keep his mouth shut.

Then Col. Tarleton began quoting from the Bible.

Hearing the Bible twisted to justify evil was more than Wilberforce could handle. Jumping up from his seat, he demanded an apology from Tarleton.

Wilberforce thundered, *"The glory of our religion is that it expressly prohibited the practice of man-stealing, calling us to act on the principle of goodwill to all."*[40]

By the time that Wilberforce had finished, the true intent of the bill was no longer a secret. As Parliament adjourned for the day, the bill became the topic of conversation for everyone. Could it really be passed? What about England's treaty with Spain that British ships would be paid to supply slaves to Spain's colonies?

How could this lucrative business be destroyed by one bill? As everyone discussed this issue, news reached England that the American Congress had just passed a bill making it illegal for any ship to bring slaves to an American port or for American ships to participate in the slave trade.[41]

While President Thomas Jefferson signed this into law in 1806, it would not be enforced until many years later when Abraham Lincoln became President and ordered his officials to prosecute slave traders.

As a young man, the first time Lincoln saw slavery, he vowed that if he ever got the chance to strike at this evil he would *"hit it hard."* The opportunity would come many years later.

Nathaniel Gordon was a well known slave trader who had escaped prosecution for years.

In 1860, Gordon was captain of a ship, transporting 890 slaves, including 600 children. When he was caught near Cuba, he was sent to New York to stand trial. He was fully prosecuted and sentenced to die under this law. Strangely, some people begged Lincoln to pardon him.

Lincoln refused and ordered Gordon to *"Relinquish all hope of pardon by human authority and refer himself alone to the mercy of God."*[42]

Gordon's execution on February 21, 1862 sent a powerful message to other slave traders still working during the Civil War.

Lincoln also reversed prior government policy by having Secretary of State Seward send a message to England that the British Navy would be allowed to stop and search American ships if there was *"reasonable suspicion"* they were slave traders.

(Now back to the story of Wilberforce.)

In England, public opinion was changing with more and more people supporting abolition.

Yet there still was not enough votes in the House of Commons to pass the bill.

So Prime Minister Grenville took matters into his own hand. Delaying the vote in the House of Commons, he made Parliament adjourn until after the election.

When the election was held, the balance of power shifted in the House of Commons. New abolition Parliament members were voted in. For the first time, the bill actually had a chance to pass.

As soon as the next Parliament was seated, Grenville introduced the bill again in the House of Lords. This time it passed by the much bigger majority of 100 to 34.

The bill went to the House of Commons and was passed by a vote of 283 to 16.

It was an unbelievable moment. Wilberforce sat in his seat with tears filling his eyes as people rushed to congratulate him on a miracle.

Passing this bill had a profound effect on the international shipping industry. It enabled the British Navy to patrol the ocean, stopping and searching any suspicious looking ship.

Thousands of slaves were freed by the British Navy.

According to *"official"* numbers from Committees of the British Government, from 1810-1849: 1,020 slave ships were seized and 116,862 slaves rescued by the British Navy.

England also developed treaties with other nations allowing ships to be stopped and searched, even if no slaves were on board, but there was reasonable suspicion it was being used for the trade.

For example, a treaty between England, Austria, Prussia, Russia and France defined reasonable suspicion as finding evidence of :

"Hatches with open gratings, instead of closed hatches used by merchant vessels,"

"Divisions or bulkheads in the hold or on deck in greater number than are necessary for ships engaged in lawful trade,"

"Spare planks fitted for being laid down as a second or slave deck,"

"Shackles, bolts, or handcuffs,"

"Large quantity of water more than needed by crew,"

"Cooking apparatus of an unusual size,"

"A quantity of mats greater than necessary for merchant vessels."[43]

This resulted in slave traders trying to conceal their activities by using smaller ships that could outmaneuver and get away before being caught. However, since many of the smaller, lighter ships had not been designed for this trade, they did not have the standard features to prevent slaves from escaping.

Thus, in several instances such as the *Amistad*, slaves were able to break free and overpower the crew.

When slave ships such as the *Enterprise* and the *Hermosa* were blown off course by bad weather and ended up in the British colonies, they were freed under British law.

In the case of the *Creole* ship, the slaves on board revolted and deliberately headed for the British territory in the Bahamas, knowing that there was freedom there.

Yet the battle was far from over. Spain, Portugal, and Brazil were still importing thousands of slaves from Africa, while the value of slaves had risen in British colonies since they could not be imported. Year after year passed as Wilberforce continued to stand up in Parliament and wonder how a free nation like England *"Should allow slavery in any place under its control?"*[44]

He was consistently opposed by other Parliament members who believed the subject should be left up to the colonies themselves.

Wilberforce responded that these colonies had not abolished slavery in the last one hundred years, why would they now? The silence was deafening. Parliament listened to his speeches but did nothing.

Wilberforce's words were heard by the slaves on the plantations thousands of miles away in the British colonies.

Throughout the colonies, news spread of his battle for abolition. When Parliament refused to do anything, the slaves revolted.

On plantation after plantation they sat down in the fields and refused to work until they received wages and fair treatment. The rebellions were brutally crushed.

When Wilberforce was blamed for causing the revolts, he responded that his only regret was not fighting harder for freedom.

In 1831, a Baptist preacher named Samuel Sharpe organized a peaceful protest in the British colony of Jamaica.

As a slave himself, Sharpe knew that the best time for a revolt was when the crops of sugar cane were ready to be harvested. If the slaves went on strike, the planters would feel the pressure to meet their demands before the sugar cane crop would be too spoiled to harvest.

Already Sharpe had a large influence on the island, serving as a deacon in the church that the slaves regularly attended as well as preaching throughout the island. His favorite sermon topic was how slavery was forbidden by the Bible. During the Christmas holidays of 1831, Sharpe led a revolt.

It was supposed to be a peaceful strike where the slaves could negotiate their freedom. But when the military troops arrived and began hurting them, the slaves responded by setting the fields on fire. Thousands of acres of sugar cane burned as the troops brutally crushed the rebellion. In the end, the harvest was destroyed, the rebellion was defeated and Samuel Sharpe was executed for treason.

The plantation owners retaliated by destroying several Baptist and Methodist churches. The remaining slaves who were pastors were forbidden from preaching. George Greathead Taylor refused and was killed, as he boldly declared, *"I can't give up serving my God."*[45]

Parliament launched a full investigation into the revolt.

The committee reported back that the only solution was abolition. Once again an abolition bill was introduced.

This time Wilberforce was not involved. At seventy-three years old, his health had deteriorated, until he could no longer continue his work. Confined to his home with barely enough energy left to walk in his garden, he had kept up with the news, hoping and praying that the battle would finally be won.

On July 26, 1833, Wilberforce received the good news. The bill to abolish slavery throughout the British Empire had just passed the House of Lords. It had more than enough votes to pass the House of Commons.

However, this wasn't the bill that he had wanted. This bill provided emancipation in exchange for compensation to the planters, which Wilberforce had always opposed.

Yet knowing there was no other way to get it through Parliament, Wilberforce was grateful for some type of victory.

Wilberforce told those closest to him, *"Thank God that I should have lived to witness a day in which England is willing to give twenty million sterling for the abolition of slavery."*[46]

Three days later, Wilberforce walked through Heaven's Pearly Gates, as across the British Empire chains were broken. Thousands of slaves walked free. Wilberforce's dream had come true.

At the news of Wilberforce's passing, Parliament adjourned so members could pay their respects. Thousands attended his memorial service. He was buried in a place of honor, close to his friend William Pitt, at Westminster Abbey.

There to honor his legacy, a statute was built, with the inscription that Wilberforce would always be remembered for leading reforms in England. *"His reliance on God was not in vain."*[47]

3

Benjamin Lay

"Let God be true and every man a liar."
Romans 3:4a (KJV)

Benjamin Lay
1682-1759
From England

In the history of the Underground Railroad, there were many brave Christians who risked their lives to help others. They believed God's command in Deuteronomy 23:15-16, *"You shall not return a slave to his master who has escaped unto you."*

While Christians from all backgrounds were heavily involved in the Underground Railroad (UGRR), the one group that stands out the most was the Quakers. These brave men and women became a huge part of the UGRR, saving many lives.

George Fox (1624-1691) founded the Quakers in England to teach people how to be led by the Holy Spirit. Emphasizing the priesthood of all believers, Quakers were known for rejecting formal church rituals for simple Bible studies.

Quakers got in trouble with the law for leaving the official Church of England. They were persecuted, thrown in prison and fined heavily.

One of those Quakers thrown in prison was a nobleman named William Penn (1644-1718). Coming from a wealthy family soon got him released from prison. Then the door opened for him to move to America.

In 1682, Penn established the colony of Pennsylvania as a safe haven for religious freedom. The local government was composed of Quakers who wrote laws protecting freedom of worship, trial by jury and a more humane criminal justice system.

In a time when British law gave the death penalty to people caught stealing food, Pennsylvania only allowed capital punishment for two crimes: murder and treason. Pennsylvania became famous for protecting human rights.

Today history remembers the Quakers for their dedication to the Underground Railroad. Yet history has forgotten there was a time when the Quakers owned slaves.

The person who transformed the Quakers into abolitionists was Benjamin Lay. Yet before Lay came on the scene, there were others who took a stand for the truth.

Benjamin Lay Portrait

In the earliest days of Pennsylvania, many of its leaders, including William Penn, owned slaves. Their hypocrisy was deeply disturbing to other people in Pennsylvania who had come to America for freedom.

In 1688, at the Quaker meeting in Germantown, Pennsylvania, a brave group of four men stood up and protested the Quaker leadership. How could they teach *"Thou shalt not steal"* while stealing the very lives of others?

After Quakers had come to America to flee religious persecution, how could they demand *"liberty of conscience"* for themselves while denying *"liberty of body"* to others?

Most importantly, how could they call themselves Christians while ignoring Christ's command to *"Do unto others as you would have them do to you?"*

Speaking loud and clear to the leaders of Pennsylvania, they noted when Jesus commanded us to treat others the way we would want to be treated, He made *"No distinction for color."*

They wanted everyone to know, *"Negroes have the right to fight for their freedom."*[48]

That protest was quickly silenced. Too much money was being made from the evil business so the leaders of Pennsylvania didn't want anyone interfering with their profits.

Things went from bad to worse. In 1698, British Parliament revoked the monopoly from the Royal African Company, opening the slave trade to all British merchants. Many new slave traders entered the market, increasing the industry and pouring huge sums of money into the American economy.

That was a huge blow to the abolitionists who were already facing tremendous opposition. With slave traders growing in power and influence, fewer and fewer people were willing to stand against them. The few that did, would become known as crazy fanatics.

One of them was a powerful judge in Boston who risked his career to stand up for the truth.

In 1700, the Chief Justice of the Massachusetts Superior Court, Samuel Sewall, published *The Selling of Joseph,* declaring slavery could not legally exist in America because according to Acts 17:26-29, people were all born free and equal.

He wrote, *"We are the offspring of God"* and *"God has made of one blood all nations."* Then he reminded everyone that God's law gave the death penalty to *"man-stealers"* in Exodus 12:16.

That Jesus had rebuked slavery when He said, *"Do unto others as you would that they should do unto you for this is the law."* (Matthew 7:12).

Jesus had also attacked the evil system that separated families when He said, *"What God has joined together let no man put asunder."* (Mark 10:9)

Justice Sewall's pamphlet had such a powerful effect in the colonies that slaveholders became concerned.

They found another judge in the Boston area to write a rebuttal. Justice John Saffin, appointed to the Court of Common Pleas, twisted Scriptures left and right, trying to deceive the people into believing that slaveholding was God's will.

Borrowing Aristotle's Greek philosophy, Saffin wrote that some people were born to rule and others were born to obey. That this was supposedly ordained by God in 1Corinthians 12:13-26. Saffin wrote, *"He sets forth the different sorts and offices of the members of the body. That they are all useful but not equal."*

Yet Justice Saffin had ulterior motives for writing this. One of his own indentured servants, Adam, had just finished his term of service and filed a lawsuit to gain his freedom. Justice Saffin tried everything to keep him in bondage, even stacking the jury to control the verdict.

After a long court battle, Adam won his freedom on appeal.

Meanwhile, Saffin's proslavery theology continued influencing many Christian denominations, enabling slaveholders to rise to the top offices, where they pulled strings to pressure the abolitionists into silence.

The abolitionists fought back. They filed lawsuits to help people gain their freedom. They petitioned the colonial legislatures. They published books and pamphlets to keep their ideas alive in public opinion.

In 1713, John Hepburn published *The Christian Defense of the Golden Rule.* He made the case that there was no excuse for the evil system. *"God has given man a free will so that he is master of his own choice whether good or evil."*

How could slavery be God's will when it forced people to violate all Ten Commandments? *"How will you answer before God's tribunal if you have lived contrary to the Gospel by enslaving Negroes?"*

He also grieved at how so many Christian groups, even the devout Anabaptists, were profiting from the evil system.

His book was read throughout the land. Some people heard his voice but many others were making too much money from the evil system to listen.

In 1718, Quaker William Burling of Pennsylvania published a pamphlet protesting the evil. He pointed out the hypocrisy of how Quakers taught pacifism. They refused to serve in the military or pay taxes for military protection. Yet they were totally contradicting themselves by supporting the violent wars of the slave trade.

He wrote, *"Shall we eat this cursed fruit? Can there be greater hypocrisy than for us as a people to refuse to bear arms or to pay them that do. Yet purchase the plunder, the captives at a great price?"*

"Thus we encourage the hellish practice of fighting, murdering, killing and robbing one another."[49]

In 1729, Quaker Ralph Sandiford published his abolition book writing, *"What comes from God—leads to God. What comes from the devil—leads to death and hell."*

Then he reminded them of God's command in Isaiah 58:6 to set the captives free or else their religion was useless. *"Preach liberty to the captive, loose the heavy burdens, let the oppressed go free and that you break every yoke."*[50]

That book got him in trouble with local authorities. They ordered him to stop circulating the book or face consequences. Yet the more the authorities tried to crush the abolition movement, the more it would grow. God would soon raise up a voice that they would not be able to silence.

LAY'S EARLY YEARS

In 1682, Benjamin Lay was born in England to a farming family. When he grew up and left home, he found a job working on a ship.

Around 1718, Lay married his sweetheart Sarah. Then they moved to the British colony of Barbados to make a living as storeowners.

There he saw something that he had never seen in England — the pure evil of slavery.

Grieved by the wickedness, he confronted the slaveholders. They mocked and ridiculed him.

His wife told him they needed to move away because she didn't want to be around these planters whose nature was *"pride and oppression."*

In 1731, Lay moved with his wife to Pennsylvania and continued the battle. At that time in American history, many Quaker leaders were slaveholders who put heavy pressure on everyone else to be silent about their evil.

This was disturbing to Benjamin Lay who couldn't understand how anyone could be a Christian without freeing their slaves. He protested by writing antislavery pamphlets. These pamphlets were read by some but mostly ignored. When no one would listen to Lay's protest against slavery, he figured out how to get their attention.

Like a John the Baptist type prophet, he lived off the grid. His food was grown on his own land because he refused to purchase food grown on plantations. He made all his own clothes. He even built his own house out of a cave on land he had bought.

At four feet, seven inches tall and humped backed, he was a powerful presence. He was known for dramatic confrontations with slaveholders.

LAY PREACHES TO PENNSYLVANIA

One day he walked several miles to town to meet with one of the town's leaders. Arriving at their home just as the family was sitting down to breakfast, he was invited to eat with them.

Lay sat down at the table. Then when the food was brought in by the family's servant, he asked, *"Is that man a slave?"*

They said yes.

Lay stood up deciding, *"Then I will not share with you the fruits of unrighteousness."*[51]

He left. Walking all those miles back home on an empty stomach. He was probably thinking of 1Cor 5:11.

One time he was walking to town when some teens on horses, rode up and started harassing him. One said mockingly to him, *"Sir, your humble servant."*

Lay replied, *"Then clean my shoes."*

The teens continued bothering him. One sarcastically asked him for the easiest way to get to Heaven.

Lay replied, *"Do justice, love mercy and walk humbly with your God."* (Micah 6:8) The teens left frustrated they couldn't get him to flinch at all.

Lay loved to read and had a large book collection at his home. He often wrote notes all over the books. In one volume he wrote a note to himself, *"A selfish spirit is satan's spirit."*

In another book he wrote, *"Cursed love of money corrupts the mind and darkens the understanding. Oh the blessed doctrine and practice of early Christians who kept out luxury, pride and cursed covetousness."*

Another time he showed up at church, while wearing very little clothing on a cold day. When people complained about his lack of clothes, he asked why they cared.

"Ah, you pretend compassion for me but you do not feel for the poor slaves in your fields who go all winter half clad."[52]

People thought he was crazy. Yet no matter how many hearts were closed to his message, Lay wouldn't stop preaching. Many times he traveled for miles to attend Quaker services. Then he sat in the back and listened to the service. After the service ended, he stood up and preached.

One time he showed up at a Quaker church in Oxford, Pennsylvania.

After the service had ended, he stood up and boldly said, *"I do not approve of all the minister has said but I didn't come here to find fault with the preaching. I came to cry aloud against your practice of slaveholding."*[53]

They didn't like that. They threw him out. Yet he kept coming back and telling them what they didn't want to hear.

Another time he showed up at a Quaker meeting with three tobacco pipes hidden under his coat. As the service was ending, he suddenly stood up and threw the pipes into the three different groups present: male ministers, female ministers and the congregation.

His point was that all three groups were equally guilty for allowing the evil to continue.

The Quakers responded by publishing a public notice in the local newspapers condemning his *"disorderly conduct."*

They wrote: *"Whereas Benjamin Lay, a person frequenting our religious meetings and pretending to be one of us, has in a disorderly manner, taken upon himself to preach amongst us."*

"We have therefore, thought fit to give public notice that we do not esteem Benjamin Lay to be a member of our religious community, but a disorderly and obstinate person. One who slights the advice of Friends, imposes on them in his preaching and disregards the peace of church."[54]

That didn't stop Lay. He showed up at the local marketplace with a bunch of teacups.

This was back in the time when manufactured goods were rare and expensive. People valued what few manufactured goods they had. Still Lay made his point, in a way they would understand.

He shattered the teacups on the ground to protest the wickedness of trading humans—made in God's image—for tea. The people watching were shocked to see him destroy valuable property in such a reckless manner. Figuring him to be insane, the young men nearby picked up Lay and carried him away before he could do any more damage.

Back home near Philadelphia, Lay had some neighbors that owned a young slave girl. He tried to reason with them about the evil of what they were doing. When they wouldn't listen, he decided to make the point a different way.

One day, he invited their six year old son over to his house, without telling the parents. All day long the parents rushed around in a panic, desperately trying to find their little child. Finally they went to Lay's house saying, *"Oh Ben! Our child is gone. He's been missing all day."*

He replied, *"Now you know the sorrow you are inflicting on another family, whose daughter is the Negro girl that you hold in slavery. Your child is safe at my home while theirs has been torn from them by greed."*[55]

In 1737, Lay went on a forty day fast. During that time, he wrote an explosive book, declaring all planters were going to hell unless they freed their slaves. Pulling no punches, he slammed them for being too *"Proud, dainty, lazy and scornful"* to work so they stole the lives of others.

Exposing their hypocrisy, he wrote how these wolves in sheep's clothing were infiltrating the church by *"Pretending to follow the pure and holy Christian religion even while spreading their filthy leprosy."*

Then he reminded them of the Bible's warning in Isaiah 9:16:

"The leaders of this people cause them to err."[56]

He wrote, *"There is a Lying Spirit in the mouths of all them that keep or trade in Slaves."*[57]

"No greater sin can hell invent than to profane and blaspheme the truth."

"Slavery is as opposite of our principles as light is to dark and Christ is to the devil."[58]

"What comes from God—leads to truth, peace, joy and Heaven. What comes from the devil leads to sorrow, misery and destruction. Is anything more devilish than slave keeping? It is the very nature of hell itself."

"As God gave his only Son that we might have life, the devil gives his child—the merchandising of souls—that whosoever believes and trades in it might have everlasting damnation. This hellish train of filthiness is the greatest sin."[59]

Next he ripped through the propaganda that had deceived people into thinking slave owners could be Christians.

Quoting from John 8:44, he wrote, *"Now if the devil was a liar and murderer, what do you think these slave merchants are?"*

Showing how the Bible condemned slavery, he pointed to Jeremiah 22:13 (ESV), *"Woe to him who builds his house by unrighteousness.........who makes his neighbor serve him for nothing and does not give him his wages."*[60]

Then he reminded them of the warning in Revelations 13:10, *"He that leads into captivity must go into captivity."*[61]

Lay pondered the words of Christ in Matthew 7:18:

"A good tree cannot bring forth evil fruit, neither can a corrupt tree bring forth good fruit."

Lay said, *"A good tree cannot bring forth such cursed evil fruit as slave trading. This practice is the worst, the greatest sin in the world."*

"But if any should say that good trees, good people could encourage this lifestyle, that doesn't make any sense."[62]

Then he wrote how the church has a responsibility to *"separate"* itself from evil. Yet instead of purging the evil from their midst—the church was actually joining forces with evil.

The very thing God had condemned in Psalm 50:18 (CEB), *"You make friends with thieves whenever you see one; you spend your time with adulterers."*

Lay wrote how that verse applied *"To men-stealers and receivers."*[63]

"Does truth lead its ministers to keep their fellow creatures in bondage? Why are ministers so angry, in a rage and such a fury when they are rebuked for their wicked sin?"

"Truth says do to others as you would have them do to you."

"Truth says to deny yourself, take up your cross and follow Christ. Truth says that whosoever is angry with their brother without just cause is in danger of judgment (Matt 5:22). Yet these false ministers are in such a fury when they are reproved for their very great iniquity."[64]

Lay wrote, *"People will slander me for saying this. Nevertheless, God is speaking to these slave keeping ministers in Psalm 50:16 (GW)."*

"But God says to wicked people, 'How dare you quote my decrees and mouth my promises!'"[65]

He continued: *"People are asleep in their sins. They don't know it because the preachers don't dare tell them. I know of no better stumbling block than our ministers and elders doing the work of the devil by keeping slaves."*

"And thereby perverting Holy Scriptures and leading people to hell by their hypocrisy and fake humility. By preaching the Gospel of liberty while holding others in bondage—one minister does more service for the devil than twenty sinners and harlots."

Then Lay referenced something Jesus had said to the churches,

"If you do not repent, I will come to you and remove your lampstand from its place."[66]

Lay wrote, *"Time for these rusty candlesticks to get moved out of their place."*[67]

In the book, he wrote how after the death of his precious wife, other Quakers had blamed him for her passing away.

"One time I was reasoning with a famous preacher R.J. at his house in Philadelphia. He said that I loved the Negroes more than my own wife."

"He accused me of persecuting the church and being the death of my wife. I will leave him to the great judge of Heaven."[68]

Then he wrote about how inadequate he felt. *"I do see myself as unfit in every way. I worry about hurting the cause of truth, which is God's cause."*

"I have been praying for many days and nights that God would raise up someone else that people would respect more than me. For I know myself to be very despised in every way by people."

"I shall leave this to the Lord. Faithfulness and obedience is required. There is no true peace without it. For I have found long ago that it's true what Jesus said, 'He that loves anything more than Me is not worthy of me.'"[69]

He described having been inspired by other abolitionists like William Burling and Ralph Sandiford. That he had been friends with Sandiford, even when many in the community pressured him to avoid talking with Sandiford.

Finally Lay closed by giving this warning: *"To all the children of light everywhere, be careful who you receive as ministers."*

"Believe not every spirit, for lying spirits may arise among you. With false visions and lying imaginations, they handle the Word of God deceitfully."

"Believe not that spirit which ministers to others what it has not learned from the Father."[70] (Lay is referencing John 5:19, where Jesus said He only did what he had learned/seen from the Father.)

Then he committed the future to God. *"Dear friends be courageous, bold and faithful. That in the light, life and power of God, you may stand in the great day of trial. So you will be preserved clean before God and be fortified by His eternal power against all the deceit, subtlety and twistings of the serpent. I charge you all in the presence of the Lord God, to walk faithfully with God who will crush the head of the serpent."*

"Truth is stronger than all the powers of hell."[71]

His book shocked Pennsylvania, resulting in the Quaker leaders publishing a public rebuke in the local newspapers. They accused him of *"Gross abuse, not only against some of their members but against the whole (Quaker) Society. We disprove of his conduct and the printing of this book."*[72]

In 1738, when Lay attended the Quaker's convention, everyone watched him closely. He walked in, carrying a large Bible, wearing a military uniform and hiding a sword under his jacket. As people stared at him, he sat down and waited quietly while the service started.

Quaker meetings followed a completely different service format than all other Christian denominations. There was no formal order of service. Everyone would sit in silence until the Holy Spirit moved on members of the congregation to stand and share something. Usually one member would speak for a few minutes and then sit down. Then other members would speak as the Spirit moved them.

This time the Holy Spirit would move in a much different way.

As the service progressed and the chance came for him to speak, Lay stood to his feet and began to preach directly to slaveholders. Telling them that their evil was as wicked in the eyes of God as murder, he threw off his coat, pulled out his sword and plunged it into his clothing where he had a hidden a pouch of berry juice that looked like blood.

The congregation shrieked in terror as the *"blood"* splattered across the audience. Lay appeared to fall dead to the ground. Had he just killed himself?

Several women fainted. When they realized he was still alive, they carried him out the door and put him on the threshold. There he lay all night, making everyone step over him to leave the meeting.

This created quite a stir in the community but wasn't enough to make them purge out the evil. After that, Lay was totally banned from Quaker meetings. Guards were posted at the door to keep Lay out. But when those doors closed to him, he found a new audience that would listen.

LAY'S INFLUENCE

There was a time in American history when the Bible was the primary textbook in public schools. Children would learn to read by reciting from the Bible in the classroom. So Lay went to the schools, believing that if the older generation was too hard hearted to change, maybe the younger generation would listen.

Lay sat and listened to the children practice reading Bible verses. Then he taught them how the Bible demanded justice for the oppressed.

One of those children would later describe, *"When the children were reading in the Bible, he would stop them and explain particular passages for their improvement. At the time we didn't think much of his anxiety for our welfare. Now sixty years later, we remember his labors and admonitions."*[73]

For the rest of his life, Lay never stopped fighting. Eventually his words were heard. Hearts began to change.

Finally, in the last moments of his life, as he lay on his deathbed, he received the news that the Quaker denomination had changed their rules. They were making small steps towards dealing with the evil in their midst. The announcement had been made that all members had to free their slaves or be expelled. Hearing it, he closed his eyes, thanked God and said, *"Now I can die in peace."*[74]

His words would live on after him, inspiring a new generation of abolitionists. One was a young schoolteacher in Philadelphia named Anthony Benezet (1713-1784).

Benezet was a devout Quaker who had emigrated from France to America in 1731. He pioneered education for women and slaves, with night classes for those who had no other option for an education. He was also a friend of Benjamin Lay. Benezet fought tirelessly for civil rights. He wrote several abolitionist books which were distributed and read around the world.

In England, John Wesley was so moved by reading Benezet's book that it inspired him to write his own abolitionist book, most of which he copied directly from Benezet. When Benezet discovered it, he was so delighted that he reprinted Wesley's book in America, knowing Wesley's name would open many doors, spreading the truth even farther.

Little did he know that this book would reach a young man in England named William Wilberforce. In the years that followed, Wilberforce directly quoted Benezet's writings during the intense debates in Parliament.

Meanwhile, Benezet created the first antislavery organization in America. Members included powerful men like Benjamin Franklin.

In 1737, Benjamin Franklin had been the young Philadelphia printer who published Lay's book. At first Franklin didn't like Lay. He thought Lay had bad breath. At the time, Franklin owned slaves. His newspaper was financed by selling ads to slave traders. Yet he still admired Lay's courage. Being a smart businessman, Franklin was willing to publish anything people would buy. So when Lay showed up with a bunch of notes, Franklin took a good look.

It was a very messy bunch of notes. Lay had written the book by scribbling down random thoughts.

Franklin looked at the pile of notes and commented there was no organization to the *"book."*

Lay said to just publish it. That made Franklin spend a lot of time reading, organizing and editing Lay's words into a readable book format. Working on Lay's book made Franklin think about things that he didn't want to think about.

Meanwhile, Lay didn't like Franklin's editorial style. After the book was published, Lay wrote a disclaimer to readers.

"There are some passages in my book that are not as well placed as they could have been. Some errors may have escaped the press. The printer was busy with other concerns. Please excuse this and remember that this was written by a poor, illiterate man."[75]

Life moved on. Years passed by. Franklin continued owning slaves even while he admired the courage and tenacity of Lay. He even kept a drawing of Lay on his wall. Sometimes Franklin would visit him.

The story is told of how one time Franklin took some leaders of Pennsylvania to visit Lay at his house. Lay made a feast for them, apologizing that he didn't have the kind of fancy food they usually ate. *"This is not the kind of food that you have at home but such as it is you are welcome to it."*[76]

In 1772, Franklin was sent on a diplomatic mission to England to represent Pennsylvania. Anthony Benezet wrote to him to ask for help in lobbying Parliament to *"Stop the dreadful slave trade."*[77]

Franklin arrived in England just as the landmark *Somerset* court decision was given to British abolitionist Granville Sharp.

Franklin befriended Sharp, putting him in contact with Benezet so they could lead the battle on both sides of the Atlantic.

Over the next several years, Sharp and Benezet wrote to each other, as Benezet lobbied the leaders of Pennsylvania to abolish slavery, and Sharp organized the team of abolitionists who helped William Wilberforce.

In 1773, while Franklin was still living in England, representing the American colonies to the British authorities, he wrote a letter to a friend, Dean Woodward, describing how: *"The desire to abolish slavery prevails in North America. Many Pennsylvanians have set their slaves at liberty and even the Virginia Assembly has petitioned the King."*

He noted the abolition petitions were routinely rejected by the King. Every time the American colonies tried to pass a law against slavery *"Laws of that kind have always been repealed."*[78]

In 1774, as the leaders of America met together for the first time, Thomas Jefferson prepared a list of grievances that the colonies had against England.

He wrote, *"These colonies greatly desire to abolish slavery. Yet all of our attempts are continually defeated by the King. He rejects the laws, we try to make, for the most trifling reasons and sometimes for no reason at all."*

"Thus he prefers to benefit a few British merchants over the lasting interests of the United States and the rights of human nature."[79]

This was still on Jefferson's mind when he wrote the first draft of the *Declaration of Independence*. In listing the complaints the colonies had, he noted how the King *"Has waged cruel war against human nature itself, violating the most sacred rights of life and liberty."*

"Determined to keep open a market where men should be bought and sold, he has suppressed every legislative attempt to prohibit or restrain this appalling commerce."[80]

Working on the committee with him was John Adams who later wrote that he was *"delighted"* with that paragraph, *"Though I knew his southern brethren would never allow it to pass in Congress."*[81]

When the *Declaration* was presented to the Continental Congress for approval, South Carolina and Georgia demanded the paragraph be deleted. So it was changed to read that the King *"Has refused to approve laws necessary for the public good."*[82]

As America broke free of England's rule and formed its very first Continental Congress, with delegates from all thirteen colonies, the door for total emancipation was opening. For the first time America could form its own laws.

One of the first resolutions Congress passed was *"We will neither import nor purchase any slave imported after December 1, 1774, after which time we will wholly discontinue the slave trade and will not hire our vessels nor sell our commodities to those who are concerned in it."*[83]

Delegates from Georgia protested this but changed their minds when Congress threatened to prevent any exports from Georgia unless they approved the resolution.

Just a few years later, in 1780, Pennsylvania passed the *Act for the Gradual Abolition of Slavery.*

Lay would have protested that law because it took too long to get things done.

This law required children born after 1780 to work for no wages until age 28. Plus it withheld liberty from anyone born before 1780. Meanwhile the battle for the soul of America would continue for more generations.

There was another Founding Father of America who was also heavily influenced by Lay.

Dr. Benjamin Rush had signed the *Declaration of Independence*. He had also served as a medical doctor in the American military and was a very active member of Benezet's abolitionist organization.

Dr. Rush had been born and raised in Philadelphia during the time Benjamin Lay was alive. To preserve his memory, in 1790, he wrote articles about how *"There was a time when every man, woman and child in Philadelphia knew the name of Benjamin Lay."*

"The success of Lay in sowing the seeds of a principle which produced a revolution in morals, commerce and government should teach the benefactors of mankind not to despair if they don't see the fruit of their benevolent propositions or undertakings during their lifetimes."

"No seed of truth ever perishes. Some seeds will produce fruit quickly. Others will take generations in growing. But they exist and bloom forever."

Those seeds were blooming as the tide of both England and America began turning towards freedom. Dr. Benjamin Rush wrote in a letter to Granville Sharp, *"Great things have come from small beginnings. Years ago Anthony Benezet stood alone in opposing slavery. Now 3/4ths of Pennsylvania cry out against it, giving me hope of seeing it abolished. Even the clergy have begun to preach against it."*[84]

The seeds sown by Benjamin Lay continued blossoming as many other people joined the fight.

4

Historical Background
Part One

As far back as archeology can research, it has found evidence of the existence of slavery. Ancient hieroglyphics vividly depict slaves in Egypt. The Egyptians were very cruel to their slaves, making life miserable for them.

One of the world's oldest known written laws, the Hammurabi Code of ancient Babylon, included harsh punishments for slaves who rebelled. Those who tried to escape would be brutally put to death along with anyone helping them.

Hammurabi's laws also allowed debt collectors to sell women and children into slavery if the man of the house couldn't pay his bills.

When the ancient Greeks brought democracy to the world, they did not bring freedom. They bought and sold white slaves.

In 324 B.C. the very first insurance contracts were invented on the Greek island of Rhodes to protect masters from the possibility of losing their slaves.[85]

The famous Greek philosopher Aristotle was a slaveowner who believed he had the right to take advantage of others.

He tried to justify his selfishness by writing, *"Humanity is divided into two groups: the masters and the slaves. The Greeks and the Barbarians. Those who have the right to command and those who are born to obey."*[86]

Denying the possibility that common people could have human rights, Aristotle reasoned, *"For that some should rule and others be ruled is a thing not only necessary but expedient. From the hour of their birth, some are marked out for subjection, others for rule."*[87]

Another powerful Greek philosopher and slaveowner, Plato, wrote, *"Nature herself intimates that it is just for the better to have more than the worse, the more powerful than the weaker, the superior ruling over and having more than the inferior."*[88]

Yet even Plato knew that the human heart yearned for freedom. Plato wrote about how since slave revolts were frequent, the best response was to avoid having slaves who spoke the same language.[89]

(Some historians today believe that America inherited ideas of democracy from these ancient Greek philosophers. Yet John Adams wrote in an 1813 letter to Thomas Jefferson that Plato's philosophy was *"absurd."*[90])

Greece also had the ancient Spartan warriors, who were known as some of the most ruthless slave owners. They had no mercy on the Helots, forcing them to do all their hardest work. Several times a year, each one was brutally whipped for no other reason than *"That they would never forget that they are slaves."*[91]

Their culture had a horrific right of passage. Spartan teenagers proved their manhood by inventing creative ways to kill slaves without being caught.

When an earthquake rocked Sparta in 464 B.C., the Helots saw their chance to revolt. Rising up to in a desperate bid for freedom, they fought back. It was a very intense battle with the Helots greatly outnumbering the Spartans. Yet the Spartans had the advantage of better military training. The rebellion was rapidly crushed.

Life got back to normal until the Persian army attacked Greece. King Darius of Persia sent messengers to Sparta demanding that they unconditionally surrender to him. King Leonidas of Sparta killed the messengers to send a clear message that there would be no surrender.

The Spartans would rather die in battle than become slaves. This led to the famous Battle of Thermopylae, which turned the tide of the war. However no sooner would peace return to Greece than civil war would rip it apart.

Fighting battle after battle, weakened the fighting forces of Sparta, giving the Helots another chance to try to gain their freedom.

In 379 B.C., the Helots set fire to Sparta and attacked. This time their revolt succeeded. Way outnumbering the Spartans, the Helots won their freedom in vicious battle. Eventually Sparta was absorbed into the rapidly growing Roman Empire.

For thousands of years, it had been considered normal around the world for the stronger to take advantage of the weaker. As Julius Caesar himself declared, *"The human race exists for the sake of a few."*[92]

Rulers of nations oppressed their own people, never considering the possibility that the common people could have human rights given to them by their Creator.

When Jesus walked the Earth, He taught a different way. Setting a powerful example, He *"came not to be served, but to serve and give His life."*

Jesus insisted that although, *"You know that the rulers in this world lord it over their people and officials flaunt their authority over those under them. But among you it will be different. Whoever wants to be a leader among you must be your servant."*[93]

This new way of thinking was completely different than the world Jesus lived in.

The Roman Empire had conquered the known world, forcing many conquered peoples of all skin colors into slavery. With far more slaves than free people, Rome had very harsh laws to keep their people under control.

When one slave killed his master for sleeping with his wife, every other slave in that household was put to death, lest anyone should follow their example.

That didn't stop the gladiator Spartacus from leading one of the largest slave revolts in history. Using his prior military experience, Spartacus led thousands of gladiators to rise up and fight their way out of captivity. They escaped, successfully fighting guerrilla warfare against the Roman army for almost two years. Several different times they defeated the full force of the Roman army.

Eventually the Roman Senate had to call for help for more troops from other areas. In the end, Spartacus died fighting in battle. The Romans crucified six thousand of his brave warriors to remind everyone that resistance was futile.

Yet slavery itself destroyed the Roman Empire. Free citizens couldn't find paying jobs. Trying to solve the problem, Roman leader Caesar wrote a law requiring at least one third of the labor had to be performed by paid workers. Nobody listened to him. Unemployed free citizens became dependent on the empire for support. Rome went bankrupt.

As the Roman Empire broke apart and history entered the Middle Ages, the system of slavery continued to thrive. England had once been part of the Roman Empire when it was conquered by Julius Caesar in 55 B.C. But as the Roman Empire declined, England was able to develop its independence.

The Anglo-Saxons, who rose to power in England, were known for running the white slave trade. Rival tribes would war against each other, with the winners enslaving the losers. Some of the Vikings attacked villages, kidnapping and selling men, women, and children.

In the sixth century, Pope Gregory described seeing young boys from England being sold in the slave markets of Rome. The city of Bristol, England even had designated slave markets to sell British children to the Irish.

The most famous white slave was St. Patrick. As a teenager in the fifth century, he was kidnapped from his home in England and sold into slavery in Ireland. For six years he was forced to work as a shepherd.

During this time he received a vision from God. He escaped from slavery and spent the rest of his life spreading the gospel throughout Ireland.

His preaching slowly began to transform the nation. Many years later, Ireland became one of the first nations to abolish slavery. After experiencing some very severe natural disasters, the people of Ireland repented of having allowed the slave trade. They destroyed the auction blocks, freed the slaves and begged God's forgiveness.

Meanwhile, the slave trade continued oppressing thousands of people around the world. Yet people always fought back.

Around the year 869 A.D., there was a violent revolt of African slaves near the Tigris River in what is present day Iraq. The Zanj was the word used to describe the people who had been forcibly taken from Africa and brought to Iraq to develop the salt marshes into plantations. This was back breaking work under brutal temperatures, with very little food or rest.

With little more than their bare hands and the absolute minimum of weaponry, the Zanj slaves managed to overthrow their masters and gain their freedom. For over a decade, they preserved their freedom, formed their own nation, and developed their own currency.

They survived as an independent civilization, until being overpowered in a vicious onslaught in 883 A.D. Many of the rebels died, fighting to their last breath. They would be remembered in history as one of the most successful slave revolts because they had survived for fifteen years. (Note: Spartacus' revolt in Rome only lasted three years).

The slave trade continued growing around the world, crushing any attempt at interference.

Throughout the Middle Ages, the Barbary Pirates from the coast of North Africa made a fortune by sailing across the Mediterranean to the coastal towns of Europe, kidnapping Europeans and selling them into slavery.

Desperate families back home tried to raise the large sums of money necessary to redeem them, but many were unsuccessful.

No village on the coast of Europe was safe as the Barbary Pirates even carried out raids as far north as England and Iceland. The raids continued many years.

Thomas Pellow of England was eleven years old when he was forced into bondage. His uncle worked on a ship so Pellow had wanted to travel on the ship with his uncle. That ship was captured by the Barbary Pirates.

The next twenty years of his life were spent in slavery in Morocco. He was trained as a soldier and forced to go on expeditions into the mainland of Africa to obtain more slaves. Eventually he found a chance to escape. He made it back home to England where he published a book detailing his experiences and the brutality of the Barbary Pirates.

When ships began transporting settlers from Europe to America, some were attacked and the passengers sold into slavery.

Massachusetts Governor John Winthrop wrote in his journal in 1638 about a family who came to America. They didn't like the frontier life so they decided to return to England. On the way back, their ship was captured and the whole family sold into slavery.[94]

Years later, when America became a nation, it lost the protection of England on the high seas. When American ships traveled to Europe, they were attacked by the Barbary Pirates, who kidnaped American sailors and demanded large sums of money in ransom.

Hundreds of thousands of dollars were paid to the Barbary Pirates by the American government. When more and more ransom money was demanded, finally President Thomas Jefferson decided to send the marines instead. The marines joined with local forces to defeat the Barbary Pirates and rescue the captives.

Now going back to the timeline of history:

In the 1400's, Portugal developed smaller, faster ships, which could maneuver better, to protect their people because their coastline was too close to the Barbary Coast and had been attacked many times by the pirates.

This new navigational technology enabled Portugal to send explorers around the world.

Hoping to find gold, or at least a new trade route away from the pirates, they explored down the coast of Africa. There they found active slave markets with traders from all over the Mediterranean making a lot of money.

In 1425, Portuguese sailors attacked slave traders from Morocco and captured their ship. On the ship were fifty-seven Africans. They were taken back to Portugal and sold. Realizing how profitable the brutal trade was, the King of Portugal formed a company to begin transporting African slaves.

Portuguese explorers conquered Brazil, setting up plantations to grow sugar, tobacco and coffee. In the years that followed, Brazil became the largest importer of African slaves.

According to historian Henry Louis Gates' study on the transatlantic slave trade, approximately 11,000,000 slaves were taken from Africa between 1501 and 1866. Most of them were taken to Brazil. 1,000,000 went to Jamaica, over 700,000 to Haiti, another 700,000 were sent to Cuba. Mexico received about 550,000, while approximately 450,000 went to America.[95]

As long as evil existed in the world, there were always people trying to destroy it. In Brazil, a Catholic priest named Antonio Vieira, decided to obey God rather than men, by telling the slaveholders in his congregation what they didn't want to hear.

In 1653, he stood in his pulpit and preached a fire and brimstone sermon, making something very clear. Unless they all repented, by freeing their slaves immediately, *"You are all going to hell where your soul will be a slave for eternity. Any man who deprives others of their freedom is condemned. God has commanded me to tell you to break the chains of injustice and let free those whom you have oppressed."*[96]

That sermon infuriated the congregation. They threw him out of Brazil and kept slavery legal in Brazil for over two hundred years. It would continue more than twenty years after it had been abolished in America.

After the Civil War, some American slaveholders moved to Brazil because it was the only way they could continue their evil system. However, things didn't work out for them, because the price of slaves in Brazil was too expensive for most of them to afford.

Many other American slaveholders moved to Cuba, including U.S. Senators Robert Toombs and John Breckinridge.

Yet they quickly realized that once America had destroyed slavery it was doomed around the world.

The slaves in Cuba were quickly becoming inspired by the *"Ideas engendered by American emancipation."*[97]

Some slaveholders like U.S. Congressman Thomas Hindmen took their slaves with them to Mexico, only to watch helplessly as the slaves quickly disappeared.

Once slavery was abolished in America, it would not last very long around the world. About twenty years after America had destroyed the evil system, Brazil officially abolished slavery in 1888.

Pope Leo XIII would send them a letter of congratulations, making it very clear that slavery was wrong according to the teachings of Jesus Christ.

Reversing hundreds of years of theological tradition, Pope Leo taught from the Bible how God made us for freedom—not the greed of other people. And in 2015, when Pope Francis visited South America, he publicly repented for what had been done to the native population.

5

Historical Background
Part Two

How did the horrible evil of slavery sink its teeth into American soil? To understand our history, we must go all the way back to the very beginning.

Spain was both the first nation to explore America and also the first to bring slavery to it. Yet the very first ship bringing slaves to America also brought an abolitionist, who inspired a revolt.

When Spain took over the island of Hispaniola (present day Haiti and Dominican Republic), it sent settlers to pioneer new communities. Among these early settlers were priests who planted churches.

Antonio Montesino was one of the very first priests who came to Hispaniola with the early settlers. However, he had his own ideas about right and wrong. Grieving over how the Spanish were enslaving the natives on the island, Montesino decided to tell his congregation what they didn't want to hear.

On December 21, 1511, he stood in his pulpit, opened his Bible and faced a congregation of slaveholders. Then he thundered from the pulpit about how much God hated slavery.

The congregation squirmed in their seats when he asked, *"Tell me by what right do you hold these Indians in such a cruel servitude?"*[98]

Then he told his congregation that unless they freed their slaves, they were all going to hell. The congregation grew enraged. Listening in the audience was the Governor of the island, and son of Christopher Columbus, Diego Columbus.

Diego was furious at this ridiculous sermon. How dare this priest insult them! Concerned about these dangerous opinions, Diego complained to King Ferdinand II of Spain.

The King responded by ordering Montesino to return to Spain before he could do any more damage. The King thought: *"Every hour that he remains in the islands holding such wrong ideas he will do much harm."*[99]

It was too late. God's Word never returns void. That sermon had touched hearts. It inspired Bartolome Casas to free his own slaves. Then he advocated for the rights of the natives. It would take years but eventually his efforts would result in changing some of Spain's laws. But when Casas lobbied Spain to stop oppressing the natives, something even worse happened.

Spain started importing African slaves to work in its colonies. In the years that followed, that island would be shaken by several slave revolts.

Meanwhile, Montesino the priest never stopped preaching. After being sent home to Spain, Montesino requested another opportunity to take the gospel to the new world. It came sooner than expected.

Montesino was invited to go to America, when explorer Lucas Ayllon received permission from the King of Spain to start the first Spanish colony on American soil.

In 1526, Ayllon took several hundred settlers, and one hundred African slaves, to start a new colony on the shores of North Carolina. This was the first time that slaves were brought to American soil.

Life on the new frontier turned out to be much more difficult than they had expected.

While they brought seeds and tools with them, they arrived too late in the season to plant crops. The entire colony had to survive on what food they had brought and whatever they could find.

At first they were able to find plenty of easy to catch fish in the river. Then winter came. Food became much harder to find.

The colony began to starve. Hunger weakened their bodies.

Illnesses spread in the colony, killing many people. Even the leader, Lucas Ayllon, got sick and passed away. The settlers became restless. Quarrels broke out, leaving the colonists at each other's throats.

Some of the colonists ran away to the nearby Native Americans who took them in and fed them. But the Native Americans did not trust them. Remembering how the Spanish ships had been sailing up and down the coast, attacking their villages and kidnapping their people into slavery, they decided to get revenge.

Late one night, after the Spanish guests had fallen asleep, the Native Americans killed everyone that had come to their camp.

Meanwhile, the rest of the settlers in the colony were still fighting among themselves. The settlement broke up into two groups. One was led by Captain Francisco Gomez who had been left in charge. The other group was led by Gines Doncel, a nobleman from Hispaniola who wanted to be in charge.

Both men declared themselves the new leader of the colony. The Doncel group defeated the Gomez group, arrested Gomez and locked him up where he couldn't cause any trouble.

The slaves saw their chance. By this time the population of the colony had dwindled significantly, leaving less people to stop an escape. They just needed to distract everyone long enough to give them time to run away.

Setting fire to Doncel's house, the slaves took off, as everyone else tried to put out the fire. In the chaos, Gomez regained control and killed Doncel. Then Gomez took the rest of the colonists back to Cuba.

No one knows what happened to the slaves after they escaped but they were never recaptured. They lived out their days in freedom in America, most likely joining forces with the Native Americans.

Meanwhile, the priest Antonio Monesino, who had fought for their rights, was one of the one hundred and fifty settlers to make it back safely to Cuba. He spent the rest of his life serving in the ministry. He died, taking the gospel to Venezuela.

After the failure of the North Carolina colony, Spain continued trying to establish a permanent settlement on American soil. Trouble continued to follow.

Pioneer life was so harsh and uncomfortable that group after group attempted to begin new communities only to give up and go back home. Not until 1565, did they succeed by discovering that life was easier in Florida where there was warmer weather.

Spanish explorer Pedro Aviles developed the successful Florida settlement of St. Augustine in 1565. He brought five hundred African slaves with him.

Many of those slaves ran away to the Native Americans. Maroon societies were formed. They lived off the land and conducted successful night raids on the plantations. While slavery continued in Florida for many years, it would be viciously attacked and eventually overthrown.

The seeds of freedom sown by Monesino continued to sprout over the next several generations. In 1835, there was a massive slave revolt in Florida. After dozens of slaves had escaped to Indian territory, they joined forces with the Native Americans and attacked the plantations. Hundreds of slaves were freed. The violence escalated until the U.S. Army was sent to restore order.

They were never able to defeat the slave revolt. After sustaining heavy losses, the army was evacuated, while Congress paid damages to the slaveholders for the loss of their *"property."*

Now going back to the beginning of America......

In 1585, England sent Sir Walter Raleigh with one hundred men to establish a permanent settlement in Virginia. Things didn't work out. Within a short period of time, everyone gave up and went back to England.

In 1587, Raleigh tried again. This time he brought one hundred and fifteen men and women, thinking that they would have a better chance.

Once again, the plan failed. The settlers disappeared without a trace. Several months later when Raleigh returned to check on the colony, he only found the word *"Croatoan"* carved into a fence post.

The first successful British colony in America was launched in 1607 at Jamestown, Virginia. However, this colony had a major problem. No one wanted to work. Born into privilege and status in England, the gentleman refused to do physical labor.

According to Captain John Smith, *"We were about two hundred people, but not even twenty workers. They had no idea what a day's work was."*[100]

This resulted in the colony eating up all the supplies they had brought with them, without planting enough food to sustain them through the next year.

When the food ran out, they tried to survive on what they could beg, borrow or steal from the Native Americans.

Within a year, most of the colony had died from starvation and disease. The rest turned to cannibalism. They barely survived, until a ship carrying supplies reached them in time before the few remaining colonists starved to death.

On that ship was John Rolfe, who had the answer to save the colony. He brought tobacco seeds that the colonists could plant, cultivate and sell at a profit.

Rolfe taught the Virginia colonists to raise it. They did under a new rule in the colony that anyone who did not work could not eat. When the first crop of tobacco was harvested, Rolfe took it back to England and sold it for a nice profit. Then he returned to Virginia and continued planting tobacco. Within a few years, they had plenty to eat and a steady income.

However, they still wanted someone else to do the work for them. Knowing that many poor people in England would want a chance for a fresh start, they started a program of indentured servitude.

People who could not afford to buy a ticket to sail to America could trade several years of labor for free passage.

At the end of their servitude, they would receive several acres of land, with seed to plant, and enough tools to farm for themselves. This opportunity attracted many people to the new world and the colony began to grow.

Then something happened in the year 1619.

From time to time, various merchant ships visited the Jamestown colony to trade with the settlers. Tobacco raised in the colony was traded to the merchant ships in exchange for food and other supplies brought by the ships.

Then the merchants transported the tobacco to Europe, or other European colonies, where it was sold at a major profit. So everything seemed normal when a ship arrived in the harbor at Jamestown in August 1619.

It was a pirate ship with stolen cargo. While the ship *White Lion* was flying a Dutch flag, it was actually manned by a British crew.

They had been stealing cargos from Spanish and Portuguese ships, who were crossing the ocean, moving plunder from colonies to the homeland.

Mostly the pirates were after loads of highly valuable gold being transported from mines in the colonies but they would take any cargo they could sell. This time, they had stolen something more valuable.

The situation had started with the Portuguese ship *San Juan Bautista* picking up three hundred and fifty slaves in Angola, Africa. They were headed to plantations in the Caribbean, when they were attacked by the British ships *Treasurer* and the *White Lion*. The British pirates took about two hundred of the slaves and sailed away.

Because the slave trade was controlled by powerful monopolies, the pirates could not just sell their stolen cargo anywhere. They had to find a place to trade it where people wouldn't care where it had come from. So they split up.

The *Treasurer* took some of the slaves and sold them in the British controlled territory of Bermuda. The *White Lion* went to Jamestown and traded twenty slaves for food and tobacco.

The leaders of Virginia assigned these twenty slaves to work as indentured servants, with the possibility of gaining their freedom after completing a term of service.

According to colonial records, in 1635, the slaves petitioned the Virginia court for their freedom. They received it. Each of them was given land and supplies to start a new life. One of them was Anthony Johnson.

Johnson had fallen in love and married one of the ladies that had come in that original group. His wife's name was Mary. Together they started a family and worked the land they had received. His farm soon prospered so much that he was able to pay to import indentured servants to work for him.

By 1651, he owned two hundred and fifty acres and a number of white and black servants. When a fire destroyed some of his property, he petitioned the Virginia court for help.

On February 28, 1652, the Virginia court decided, *"Anthony and Mary Johnson have resided in Virginia for over thirty years. Considering their hard labor and honored service in this county and the great losses they sustained by an unfortunate fire, it is ordered that for the rest of their lives, they and their two daughters are free from paying taxes in Northampton County."*[101]

Johnson was living the American dream. Yet having a beautiful family, nice home, successful farming business and tax freedom wasn't enough for him. Even though as a former slave he knew how precious freedom was, he refused to give it to his own servants.

In 1653, when one of his own indentured servants completed his term and petitioned for freedom, Johnson fought to stop it.

That servant, John Casor, ran away to the neighboring farm owned by Robert Parker.

Then Casor filed a lawsuit claiming his freedom. *"I came to Virginia with a contract for seven or eight years of indenture. Anthony Johnson has kept me seven years longer."*

The court ordered Johnson to produce the contract of indenture. He denied that it existed, claiming, *"I own this negro for life."*

Witnesses were called. Johnson's neighbors Robert and George Parker testified in court, *"We know this Negro. He had a contract."* They confirmed that John Casor deserved his freedom and unpaid wages.

Anthony Johnson fought back. Filing a counter lawsuit against Robert Parker, Johnson complained that Parker was illegally assisting a runaway slave. *"Robert Parker is detaining my servant John Casor on the pretense that Casor is a freeman."*[102]

The court agreed. Johnson won the case. Casor was sentenced to perpetual slavery and Robert Parker was ordered to pay all court costs for the lawsuit.

This horrible court decision was possible because the Virginia court was controlled by rich planters. They didn't care about justice. They only cared about making money. They wanted to force many more indentured servants to be held in bondage way beyond their contracted times. The abuse would continue with more horrific court decisions trampling on the rights of others.

In 1641, a white indentured servant named William Andrews was sentenced to lifelong slavery for physically attacking his master. Two more examples are John Haslewood and Giles Player who received the same punishment for theft.[103]

Meanwhile, Anthony Johnson became the first of many African American slave owners, who would continue profiting on the slave trade for the next two hundred years until the Civil War brought freedom.

From just after the Revolutionary War in 1790 to the Civil War in 1860, every ten years the U.S. Census data recorded significant numbers of African American slave owners.

(There are also other records such as court proceedings and tax records.) For example, in 1864, the city of Charleston, South Carolina had eighty-one black slave owners, who owned a combined total of two hundred forty-one slaves.

Some of them had once been slaves themselves but found a way to move up to owning their own slaves. One example was Sally Seymour who was freed by her master Thomas Martin in 1795. She started a successful bakery, purchasing four slaves to work for her. When she passed away, she left her slaves as an inheritance to her children.[104]

William Ellison was another freed slave who became a slave owner himself. He worked his way up from being a repairman of cotton gins to buying his own plantation and slaves.

By 1859, he owned thirty-five slaves and two hundred acres of land. His crop harvest produced eighty bales of cotton, which sold for $3,840.[105] He continued reinvesting his profits until by the time of the Civil War, he owned sixty-three slaves.

The wealthiest slave owner in the state of South Carolina was James Pendarvis. Born from a relationship between a planter and a slave, Pendarvis would inherit most of his father's plantation and slaves.

When Pendarvis himself died in 1798, he left an estate to his own children of 5,987 pounds sterling, 3,000 acres of land, 155 slaves, 99 head of cattle and 107 oxen.[106]

When the Civil War came, many black slave owners anticipated the potential loss of their *"assets"* and diversified their holdings by investing in real estate. They would make a nice profit from renting out their real estate.

Anthony Weston lost an estate worth $12,600 when his fourteen slaves were freed after the Civil War. He survived by renting his real estate holdings valued at $30,000 in 1870.[107]

The evil system made lots of money until it was destroyed by the sacrifice of hundreds of thousands of American soldiers giving their lives on the bloody battlefields of the Civil War.

6

Historical Background
Part Three

From the very beginning of slavery in America—it was always about the money.

In 1619, British pirates brought slaves to Virginia because the love of money is the root of all evil.

In 1620, the Mayflower landed, bringing British settlers to the New England area. That group actually did come here to worship God. Yet soon after many other ships arrived, bringing many more settlers. Some of these settlers just wanted a fresh start and were willing to work with their hands. Others wanted someone else to do the work for them.

In 1624, Samuel Maverick, a British settler, moved to America bringing two slaves with him.[108] He appears to have been the first slaveholder in the Boston area and owned white indentured servants.

In 1638, British attorney Emmanuel Downing, came to Massachusetts. After trying to make a living for several years, in 1645, he wrote to the Governor of the colony, John Winthrop, and complained about how expensive it was to pay for labor. He didn't like how his indentured servants were eager to leave and start their own lives after completing their term, instead of staying and working for very low wages.

Downing wrote, *"I do not see how we can thrive until we get into a sufficient stock of slaves to do all our business."*[109]

Other people agreed with him and began looking for ways to profit from the worldwide slave trade. By then the New England colonies were producing goods like tobacco, lumber, and corn, which could be transported across the ocean and traded for slaves at other British colonies in the Caribbean. Then the slaves were brought back to New England and sold as the triangle of the transatlantic slave trade was born.

Eventually sugar produced in the Caribbean was taken to New England and manufactured into rum, which was transported to Europe and traded for various products, which were then taken to Africa and traded for slaves. Soon American ships were sailing directly from New England ports to Africa as international trade grew. The rum distilled in Massachusetts became the backbone of the slave trade.[110]

The evil continued to grow as the lust for power and money knew no bounds. On February 28, 1638, Massachusetts Governor John Winthrop recorded in his diary, *"Mr. Pierce, in the Salem ship, Desire, returned from the West Indies after seven months. He brought some cotton, tobacco and negroes."*[111]

That was only part of the story. There had been a war between British settlers and the Pequot tribe of Native Americans. The settlers had taken a number of prisoners and tried to enslave them.

When the prisoners fought back, the settlers decided to get rid of them by shipping them to the West Indies.

That was the real reason Mr. Pierce had gone to the West Indies. Then he had brought people to America against their will and sold them for lots of money.

Those horrible sins were not forgotten.

Over one hundred years later during the Revolutionary War, when the Boston harbor was blockaded by the British and the American people were suffering, some saw this as the judgment of God for the sins that Boston had committed in the slave trade.

Deacon Benjamin Coleman petitioned the Massachusetts court to overturn the oppressive laws. He wrote, *"As I have said before and will say again—our transgressions are multiplied. Wasn't Boston the first port on this continent that began the slave trade? Aren't they the first shut up by an oppressive act and brought almost to desolation?"*

"I beseech this town, by all the love you have for it, to exert yourselves to obtain a discharge for the slaves from their bondage. If this is done, deliverance (from the war) will come to us. But if this oppression is continued, my soul will weep for that crime."[112]

Only a few years later, Massachusetts became one of the first states to abolish slavery. It came as slaves heard the *Declaration of Independence* and went to court to enforce their rights.

In 1769, Founding Father John Adams used his legal knowledge as an attorney to help a slave win his freedom in a courtroom. He wrote in his journal that there were many freedom suits happening in the nation.

Then John Adams wrote the Massachusetts Constitution to protect the rights of the people. What he had written would later influence the Founding Fathers when the U.S. Constitution was written.

In 1780, the Massachusetts Constitution became the law of the land, opening the door to freedom for many.

In 1781, Elizabeth Freeman filed her lawsuit in Massachusetts and won her freedom.

In 1781, Quock Walker filed a lawsuit for his freedom. He had escaped from slavery only to be dragged back and brutally punished. He sued for damages on the grounds that slavery couldn't legally exist in America. The court found that slavery had never been *"Expressly enacted or established"* by any law. Instead it had been *"The practice of some European nations."*

Thus it could not legally exist in America because *"Whatever sentiments had formerly prevailed or slid in upon us by the example of others, (now) a different idea has taken the people of America."*

"Every person is entitled to liberty, life and property. This being the case, the idea of slavery is inconsistent with the (Massachusetts) Constitution. There can be no such thing as perpetual servitude."[113]

That was a landmark court decision in a time when much of the rest of the world felt differently. The Massachusetts Constitution again saved lives when evil people came to the state.

In 1788, slave traders kidnapped three free black citizens from Boston, put them on a ship and took them to the Caribbean to sell them into slavery. There was a huge public outcry against this evil. Boston leaders wrote letters to Caribbean authorities. When their ship landed in the Caribbean, they were rescued by local authorities and sent back to Boston. They returned home to a big party, with lots of people celebrating their release.

In the years that followed, kidnappers and slave catchers had a hard time doing dirty deeds in Boston because the public rose up and stopped them.

The spirit of freedom was growing in America. Also during that time, on September 13, 1776, we find something interesting in the records for the House of Representatives.

There's a resolution to prevent the sale of two slaves at auction. The resolution declares: *"The selling and enslaving the human species is a direct violation of the natural rights vested alike in all men by their Creator, and utterly inconsistent with the principles on which the United States has carried its struggle for liberty. Therefore all persons connected with the said negroes are forbidden to sell them."[114]*

In 1804, a Vermont Judge made a landmark court ruling in a case involving a slave escaping from New York. The master had hunted him down, went to court, showed legal title to him and demanded the court return him to bondage. The Judge asked how did they obtain title for a human being? The master produced a bill of sale. The Judge replied, *"Unless you can produce a bill of sale from Almighty God, nothing else will be regarded as evidence of title to a man and your case will be dismissed."*

That was the end of slavery in Vermont.

Now going back to the timeline of history:

By 1645, shipbuilding had become an industry in New England. The first ship built was the *Rainbowe,* which was utilized to transport slaves to America.[115]

In 1646, two men from Massachusetts, William Smith and Thomas Keyser, sailed to Africa. Joining other slave traders, they attacked a village and carried away one hundred people. After dividing the captives among all the traders, Smith and Keyser returned to Massachusetts to sell their cargo.

They were in for a big surprise. Upon arrival, they were arrested and put on trial for the crime of man-stealing. They were also charged with breaking the Sabbath, since the raid had happened on a Sunday.

On November 4, 1646, the Massachusetts court ruled: *"This is our first opportunity to protest against the terrible sin of man-stealing."*

The court freed the slaves and transported them back home at the expense of the colony, *"To rectify what has been done unlawfully and to deter all others from this vile trade which is abhorred by all just men."*[116]

That Massachusetts court also prosecuted Mr. Williams for buying a slave that had been *"Stolen from Africa."*[117]

There were early antislavery laws on the books in America.

In 1642, the colony of Connecticut recorded a law giving slave traders the death penalty.

The original text of their law actually quoted from the Bible. *"If any man steals a man or mankind he shall be put to death: Exodus 21:16."*[118]

No other details were provided.

In 1652, Rhode Island also had a law forbidding slavery.

"To prevent the practice of buying Negroes for service or slaves forever, let it be ordered that no black or white person be forced to serve longer than ten years. At the end of ten years they will be set free as the English servants are."

Anyone who refused to obey the law or tried *"To sell them away that they may be enslaved to others"* would be heavily fined.[119]

Meanwhile, the struggle between freedom and slavery in America was just beginning. The massive amounts of money made in the trade bought politicians, courts and judges. Law enforcement looked the other way. No law could stop the greed of certain people determined to have their own way.

Yet even with injustice in the legal system, in the earliest years of America there were a remarkable number of slaves who successfully filed lawsuits to gain their freedom.

In 1655, Virginia resident Elizabeth Key won a lawsuit to gain her freedom. Her father was a white plantation owner named Thomas Key, who was one of the earliest settlers of Virginia. Her mother was of African descent.

When Elizabeth was born, Thomas had tried to deny he was the father. The matter was heard in a local court. Witnesses were called to the court. They testified that they had often seen her mother and father sleeping together. Her father was ordered to pay a fine for having a baby out of wedlock. When he passed away shortly thereafter, six year old Elizabeth was left in the care of a family friend. However, that family friend decided to return to England and turned Elizabeth over to a planter.

For years Elizabeth was forced to work as an indentured servant. She grew up and fell in love with another indentured servant named William Grinstead. He wanted to marry Elizabeth but couldn't since the law required illegitimate children to serve a contract of apprentice, until they were grown adults who could provide for themselves. Fortunately, while Grinstead had come to America as an indentured servant, he also was an attorney who knew the law.

He filed a lawsuit, declaring Elizabeth was free. When they lost the case, they appealed it all the way to the Virginia General Assembly.

The court ruled that servants had to be freed after they completed their term of indenture.

After winning the case, Elizabeth and William were married the following year, as soon as William had completed his own indenture.

In 1673, plantation owner George Light was sued for unlawfully detaining a black indentured servant after he had completed his five year term.

The court ordered that the servant should *"Be free from his master and that Mr. Light pay him corn, clothes, and four hundred pounds tobacco for his service according to the custom of the country."*[120]

While some people gained their freedom, others were forced into bondage. The worldwide system of slavery was still growing.

In 1672, King Charles II of England granted a monopoly on the slave trade to the Royal African Company.

In 1698, Parliament opened the slave trade to other merchants outside the monopoly. The profits were too large to allow anyone to interfere with it.

As the Earl of Dartmouth described, *"We cannot allow the colonies to limit or discourage in any manner a traffic so beneficial to the nation."*[121]

When the colony of Virginia tried to pass laws limiting the industry they were quickly voided for being *"too inconvenient."* The Virginia legislature responded by petitioning the King. *"The importation of slaves into the colonies has long been considered very inhumane. We humbly beseech your Majesty to allow the colony to permit laws that would limit this very harmful commerce."*[122]

The King ignored their pleas but the protests would continue.

As America became a nation, the forces of freedom and slavery continued to struggle for control.

This was the world that Robert Carter III was born into. This was the nation that he loved but the society that he would defy. Taking a different pathway than his family, friends and community, he would accomplish what many other people thought was impossible.

7

Robert Carter III

"And be not conformed to this world:
but be ye transformed by the renewing of your mind,
that ye may prove what is that good,
and acceptable, and perfect, will of God."
Romans 12:2 (KJV)

Robert Carter III
1727-1804
From Virginia

In 1727, Robert Carter III was born into one of the most powerful families in Virginia.

His great-grandfather, John Carter, had left England to come to America in 1635 with little more than the clothes on his back. Yet he worked his way up to accumulating several thousand acres of land and dozens of indentured servants to work it. When he passed away, the inheritance of land and servants passed down to his children who continued building the family wealth.

John Carter's son, Robert Carter I, accumulated over a hundred thousand acres in Virginia, becoming so powerful that he was able to successfully challenge the King of England in court.

Filing lawsuit after lawsuit, he fought for land rights in America. He got the land he wanted, along with many other things, like getting the Governor of Virginia dismissed when he didn't like his policies.

He was a cruel man, who cared nothing about the law. When his indentured servants appealed for their freedom at the end of their term of service, he petitioned the court to invalidate their contract, since it had not been signed before a justice of the peace. The court agreed and ordered those servants back to his plantation for life.

He also became a very profitable slave trader, importing men and women to work his plantations. However that decision would cost his life.

According to the official story, he died from a disease that might have been contracted through his habit of boarding slave ships docked in the harbor to *"inspect"* the people before purchasing.

His son, Robert Carter II, passed away from a drug overdose. The massive inheritance flowed directly down to his grandson, Robert Carter III.

Robert Carter III grew up in the lap of luxury, having anything that he wanted. Yet the only thing he really wanted was to please God's heart. In the years that followed, he would offend his friends, alienate his family and sacrifice a fortune just to do what God asked him to do.

Carter was young when his father and grandfather passed away, giving him full control of his inheritance.

His family had tried to prepare him for this day by sending him to study at the top schools. Carter had attended law school in England but did too much partying to get any studying done.

He returned home to America, without a law degree, but he would never need to work anyway. Just collecting the rents from his massive property holdings was more than enough money.

With lots of time on his hands, he tried to manage his estate. He focused on planting tobacco and collecting rents. He also loaned money to many friends including Thomas Jefferson and Benjamin Harrison.

Even the British military commander Lord Dunmore borrowed money from him. That created an awkward situation when the Revolutionary War broke out and Dunmore led the British forces attacking America. However, as a true gentleman, after the war Dunmore repaid the loan to Carter.

In his free time, which was all of his time, Carter moved in the highest social circles of Virginia. There he met and fell in love with a beautiful young lady from Maryland named Frances Tasker. Her family had made their fortune in the iron industry. They married and raised ten children together.

Yet no matter how many things filled his life, his heart yearned to know God. He often attended church. Yet while he attended the same prestigious church as George Washington, he soon tired of it.

Wealthy families in Virginia attended the local Anglican Church, which followed all the doctrine of the Church of England and used the Book of Common Prayers for ceremonies. Only pastors licensed by church authorities were allowed to preach. No preaching was allowed outside the church.

At that time, Virginia did not allow other Christian denominations. Virginia had laws against Puritans, Quakers, and Baptists. Residents of Virginia were also required to pay taxes that funded the Anglican Church.

People who missed church were fined under Virginia law, which made church attendance mandatory. Yet that didn't stop Carter from leaving the church. Secretly he had begun attending forbidden church services held by the radical new group known as Baptists.

In 1609, Anglican Pastor John Smyth left the Church of England and started the Baptist movement. Smyth preached that true worship came from the heart, not through performing boring religious rituals.

Emphasizing reading the Bible and being led by the Holy Spirit, Smyth also rejected using the Book of Common Prayer in church services.

While his movement started in England and spread to Holland, it soon reached America.

Baptist preachers traveled from town to town, preaching the gospel outdoors because no churches would allow them to speak. They were arrested, fined heavily, publicly whipped and thrown out of town after town but they kept preaching.

In 1768, three Baptist ministers were put on trial for preaching without a license in Spottsylvania County. As the charges against them were being read, the door to the courtroom opened.

Patrick Henry, who had just ridden horseback furiously for sixty miles, arrived in time to volunteer as their defense attorney.

He made a resounding speech that brought the courtroom to its feet as he thundered, *"Am I reading these charges correctly? Are these men to be tried for the crime of preaching the gospel?"*

"Yet our fathers left England for the freedom to worship God according to the Bible. Have all their sacrifices been in vain?"[123]

The whole courtroom was so moved that the judge immediately released the ministers.

Patrick Henry wasn't the only person helping these ministers.

Robert Carter also attended their services and supported them financially. He allowed some of them to live rent free and preach revivals on his property. He paid the salaries for others so they could stay on the road preaching the gospel. And Carter himself would travel as much as one hundred miles just to hear them preach.

Another Virginian helping these ministers was future President James Madison. He had noticed that the colony of Pennsylvania, founded by Quakers who allowed religious freedom, was thriving.

People were moving to Pennsylvania while Virginia was struggling to attract new settlers.

Madison lobbied the Virginia legislature to pass a law protecting religious liberty. He believed that if people had the freedom to worship as they pleased, Christianity would thrive.

Madison wrote that Jesus had succeeded without the support of the civil authorities. Christianity had thrived even under persecution in the days of the Roman Empire.

Yet Virginia's religious laws were corrupting the clergy by giving them too much power, when no one could be trusted with too much power. Bad behavior among the clergy had been so rampant that already Virginia had been forced to issue laws forbidding the clergy from drunkenness, laziness and gambling.

Eventually Madison's efforts resulted in the Virginia legislature passing one of the earliest religious liberty laws in America. The new law overturned the previous laws requiring church attendance and forcible tithing.

The text read: *"How we fulfill our duty to God can only be by conscience, not by force or violence. Therefore all men are entitled to the free exercise of religion."*[124]

But first those rights would have to be won in the Revolutionary War. When war came to America, it would disrupt life for everyone.

Carter was caught in the middle of it. As the war disrupted most business and international trade, people could no longer afford to pay their rents. The ones that still had some money, tried to pay their rents in the new American currency.

Not knowing the future outcome of the war, many merchants refused to accept American currency, demanding to be paid in gold or silver. They were worried that if England won the war, American currency would be worthless.

Carter was one of the few landlords who believed in America enough to accept rent payment in American dollars.

Carter also served his country by changing his farming practices. He knew that winning the war would require America to grow and build what it needed, instead of relying on foreign imports.

He could have made a lot of money growing tobacco. Instead, he planted food crops like potatoes and corn. Then he built a factory on his property where other crops he raised could be made into clothing.

When America's military ran short on supplies, Carter had the blacksmiths, who worked for him, begin producing bayonets.

When the military ran out of food and took his cattle to feed the troops, Carter allowed it. While he paid a heavy price for serving his country, he sacrificed, because he loved America.

That decision cost him. Soon Carter was running short on funds for several reasons, including that many of his tenants couldn't afford to pay their rents.

Carter didn't have the heart to evict them. He tried to forget about his financial problems while he attended many different types of church services including Quaker and Episcopal.

As the Revolutionary War progressed, the tide of public opinion in America turned towards freedom. Everyone was talking about Thomas Paine's book *Common Sense,* which made it very clear that God had never intended his people to be ruled by a tyrant.

Carter noted in his journal that in America it had become *"Sinful to acknowledge any being as King but the Lord."*

Carter committed his life to Christ. He wrote in a letter to Thomas Jefferson, *"I do testify that Jesus Christ is the Son of God; that through Him mankind can be saved only."*[125]

As the American Revolution continued and people discussed freedom, the northern colonies began to abolish slavery.

In 1777, Vermont became the first colony to abolish it.

In 1781, a slave living in Massachusetts, Elizabeth Freeman, heard the *Declaration of Independence* read. It changed her life. She went to a local attorney and asked, *"I heard that paper read yesterday that says that all men are born equal. That every man has a right to freedom. I'm not a dumb critter. Won't the law give me my freedom?"*[126]

The lawyer agreed and filed a freedom lawsuit for her.

This was a difficult lawsuit because her master, John Ashley, was a very powerful man. How could she succeed against a former judge and Colonel in the American military?

Yet she won the case. The court declared that she was free and ordered Colonel Ashley to pay her thirty shillings in damages.

The lawyer who helped her win the case was named Theodore Sedgwick. He hired her to work for him. She worked that paying job for many years. She also fell in love and got married. Her husband would give his life, serving in the military during the Revolutionary War.

She lived in her own house for many years, finally dying in freedom on Dec 28, 1829, at the healthy old age of eighty-five.

In 1787, Massachusetts had another successful freedom lawsuit. John Quincy Adams would later describe, *"There was a (mortgage) note given for the price of a slave in 1787. There was a lawsuit over the note and the Court ruled that the maker had received no consideration—as man could not be sold. From that time forward, slavery died in the Old Bay State."*[127]

Around that time, another slave, Quock Walker, also sued for freedom and received it, along with fifty dollars in damages when the Massachusetts State Supreme Court declared that slavery was abolished by the *"free and equal"* clause of the Massachusetts Constitution.

That clause had been written by John Adams and had gone into effect on October 25, 1780. By the time that the first U.S. Census was taken in 1790, it recorded that there were no slaves left in Massachusetts.

Meanwhile, in the southern colonies, the abolition movement was also growing. Some southerners freed their slaves during this time. One was Founding Father Robert Pleasants (1723-1801), who led the abolition movement in Virginia.

At the time, Virginia had a law forbidding anyone from freeing their slaves. Pleasants disobeyed that law by freeing his slaves, paying them wages and providing for their education.

Then he aggressively lobbied the Virginia legislature to overturn the law forbidding emancipation. He would win this battle in 1782, when Virginia overturned the anti-manumission colonial law of 1691.

Later, in a landmark court decision in 1799, Pleasants successfully sued his own family members, forcing them to free their slaves under the terms of his father's will. Pleasants continued filing many more freedom lawsuits, trying to help as many people as he could. Yet the more success he had, the more opposition he faced.

When he founded the first southern abolitionist organization and aggressively pursued freedom, the slaveholders fought back.

By 1785, they had formed their own proslavery group and began aggressively resisting any interference with their lifestyle.

They pulled every string possible—causing laws to be passed in Virginia—making it illegal for abolitionists to serve on the jury in a freedom lawsuit. Harsh penalties were also imposed on anyone filing a freedom suit. As pressure began to rise against freedom, Pleasants found his support slipping.

Feeling like he had hit a brick wall, Pleasants began writing to the leaders of Virginia, pleading with them to join the antislavery movement.

On December 11, 1785, he wrote a letter to George Washington, reminding him that as one of the most powerful men in America, God had raised him up to protect human rights.

After leading a nation into freedom, how could he withhold it from his own household? Pulling no punches, Pleasants directly asked Washington to free all his slaves to:

"Set a powerful example that would lead to universal emancipation. This is the sacrifice that God is requiring of your generation."[128]

Washington never replied to that letter. He didn't want to think about it. He didn't want anyone to know that he was busy hiring slave catchers to track down runaways from his plantation.

He hoped no one would know that days before he had left to fight the Revolutionary War, he had posted a reward for the recapture of ten runaways from his plantation: eight white and two black.

Pleasants wrote letters to many other Founding Fathers, including Thomas Jefferson, Patrick Henry, and James Madison but his voice would not be heard.

Patrick Henry wrote back to thank Pleasants for sending Anthony Benezet's book. He even admitted that he was ashamed of owning slaves. *"I will not, I cannot justify it."*[129] Yet he would try to justify it by failing to do anything about it.

This lack of support from the leaders of Virginia caused Pleasant's southern abolitionist group to soon die out.

Shortly thereafter Pleasants passed away. Meanwhile, many other abolitionists would continue the fight.

Warner Mifflin (1745-1798) was born into a Quaker slaveholding family in Virginia. When he grew up and received his inheritance, he moved with his slaves to Delaware. For several years he ran a plantation.

Then the American Revolution changed his heart into freeing his slaves. He led the abolition movement in Delaware, becoming a justice of the peace, who used his judicial powers to help slaves win freedom lawsuits. He was one of many who were deeply grieved when the principles of the *Declaration* failed to demolish slavery.

Trying to change that he visited George Washington to plead with him for social justice. Washington told Mifflin that he didn't want to talk about it.

When Washington did nothing, Mifflin wrote letters, pleading with him to change his mind, *"Don't you realize that there is a God of justice who will judge everyone according to what they have done?"*

Mifflin: *"How do you think God feels when He sees the founders of America ignoring the suffering of their African brethren?"*

"I worry for my country that we have broken the covenant we made with God when we declared that all men are created equal."[130]

When Washington ignored that letter, Mifflin petitioned Congress. Yet the Congress who had proposed abolishing slavery before the Revolutionary War, changed its mind after the war.

Congress was so offended at Mifflin's petition that they voted to return it to him to protest he had wasted their time.

Congress wanted all the other abolitionists in the nation to know *"That they will never have their way with Congress."*[131]

The battle between freedom and slavery would continue when the Founding Fathers of America met together to write the Constitution in the summer of 1787.

When America had declared independence, no one really knew how government by consent of the governed was supposed to work. How could they maintain law and order without a King to enforce it?

During the war, Congress had drafted the *Articles of Confederation* as the first type of Constitution to govern the new nation. Within the first few years of living under it, the states realized that there were too many problems.

One of the biggest problems was that during the Revolutionary War, the military constantly lacked proper supplies and equipment, because Congress could not raise funds through taxes.

Congress could only politely ask the states for funds, which of course none of them wanted to provide. Then there was also the issue of how each state should be represented in Congress. Should they favor the smaller states by allowing each state only one vote? Or should they favor the larger states by determining seats in Congress according to population?

There were lots of other issues to address, everything from how to create a court system to international trade. Most importantly, how could the power of the federal government be divided equally so that no one person could seize absolute power?

This was a powerful moment in history. The world was watching. Many people thought that democracy was impossible. That America would self-destruct in a few years.

Yet the Founding Fathers were determined to lay a foundation that would last forever.

SLAVERY DEBATED WHEN THE CONSTITUTION WAS WRITTEN

For weeks, they debated various ideas. Everyone had their own opinion. Their speeches went on for hours. James Madison, who was in charge of recording the discussions, did his best to keep up.

This was uncharted territory. They had to figure out something that had never been done before. How could government by consent of the governed really work?

As they sat down and debated everything, they ran into a major roadblock.

How was Congress going to work? The smaller states wanted each state to have an equal number of votes. The bigger states wanted to have representation determined by population. The more people living in that state, the more reps they would have in Congress. This was a very sensitive issue, since the more reps in Congress each state had, the more votes and power they had.

Congress deadlocked over this issue. As both sides dug their heels in, Benjamin Franklin tried to talk some sense into them.

Reminding them of how during the darkest days of the American Revolution when no one knew what would happen, *"We had daily prayers in this room for divine protection! Our prayers were heard and graciously answered."*

Looking around the room, Franklin asked, *"Have we now forgotten that powerful Friend? I have lived a long time and the longer I live the more convincing proofs I see that God governs in the affairs of men."*

"And if a sparrow cannot fall to the ground without his notice, can an empire rise without His aid? I believe that unless the Lord build the house, they labor in vain that build it (Psalm 127:1)."

"We seem to understand our need for wisdom because we keep running around, trying to find it. We've gone back to ancient history for models of government and examined the different republics which no longer exist."

"Now while we grope in the dark for political truth, why haven't we thought of asking the Father of lights to illuminate our understanding?"

"I believe that without His help we will end up like the Tower of Babel. Then humanity will despair of establishing government by human wisdom and leave it to chance, war, and conquest."

"Therefore I move that henceforth prayers imploring Heaven be held in this assembly every morning before we proceed to business. And that one or more of the clergy in this city be invited to officiate."[132]

Franklin sat down as the room glared back at him. Franklin had diplomatically told them that they had no idea what they were doing. Yet his proposal was rejected for fear that the people of America might find out that Congress couldn't get along.

The arguing continued, while Franklin wrote in his journal, *"Except for three or four persons, the convention thought prayers unnecessary."*[133]

This was finally settled in the famous compromise of creating two houses of Congress. In the Senate, each state has two votes. In the House of Representatives, the number of votes is determined by population.

That brought up the question of slavery. How would they count the population? Would they include only the free citizens or the slaves?

This decision would affect the balance of power.

The proposal was made that they should count all the free citizens and 3/5ths of the slaves. That sparked a furious debate on slavery with many upset about this proposal.

Luther Martin of Maryland demanded the slave trade be destroyed because *"It was inconsistent with the principles of the revolution and dishonorable to America."* He refused to sign the Constitution, storming out of the convention in protest that they had not purged out the evil.

Rufus King of Massachusetts agreed, describing how slavery was unconstitutional since it interfered with the Constitution's purpose to protect the nation from danger. By bringing the potential for a violent slave revolt, the evil system was increasing the cost of law enforcement and expenses of the federal government, while also depleting funds needed by the free states for other things.

Protesting that the tax revenue of free states would be used to protect slavery he thundered, *"How can you force one part of America to defend another part, while that other part is free to increase their danger by importing more slaves?"*

Gouverneur Morris of Pennsylvania stood up and also made it very clear that he hated slavery. *"It was a curse to any nation."*

He told everyone to travel through the United States and see for themselves the difference between the thriving economies of the free states contrasted with the *"misery and poverty"* in the slave states.

America had already proven that freedom was good for the economy because every state that had abolished slavery had seen an improvement in their local economy.

Then Morris attacked the horrible idea of the 3/5ths rule. This opened the door for the proslavery part of the nation to control the entire federal government.

By including the slaves in the population counted, then the slave states would increase their reps in Congress and their electoral votes which elected the President. They would end up with a majority that could control Congress, pass all the laws they wanted and keep the President accountable only to them as he appointed the Supreme Court. Couldn't everyone see the secret conspiracy behind it?

If that wasn't bad enough, the 3/5ths rule was also pure taxation without representation. How could they allow the slave states to increase their votes in Congress by counting the slaves, without allowing the slaves to vote? Since the battle cry of the American Revolution had been no taxation without representation, he demanded to know how slaves would be represented in Congress.

"Are they men? Then make them citizens and let them vote! Are they property? Then why is no other property included like the houses of Philadelphia?"

"The real answer is that the slaveholders of Georgia and South Carolina, who are defying the most sacred laws of humanity, shall have more votes in a government created for the protection of human rights, than the citizens of Pennsylvania or New Jersey who hate slavery."

"And what reward does the North get for sacrificing the principles of justice and humanity? Someday we will be forced to send our soldiers to defend the southern states from the very slaves that they complain about."

This infuriated southern reps like John Rutledge of South Carolina. He fired back, *"Religion and humanity have nothing to do with this issue. The only thing that matters is whether or not the South will join the Union. We will never give up slavery."*

Charles Pinckney, also from South Carolina, agreed emphasizing, *"South Carolina will never receive this plan if it prohibits the slave trade."*

The debate dragged on and on with both sides moving farther apart. Attempting to reconcile them, Oliver Ellsworth of Connecticut stood up to make a point. *"Slavery has already been abolished in Massachusetts and it's in the process of being abolished in Connecticut. Soon it will not even be a speck in our country."*

George Mason of Virginia gave a long speech about how slavery hurt the nation's economy by destroying the work ethic and *"Discouraging the arts and manufacturing."*

He blamed the British merchants for bringing slavery to America in the first place and the British government for *"Constantly stopping any attempt to abolish."*

Now America had to face the reality that *"Slavery will bring the judgment of Heaven on a country because national sins will be punished by national calamities."*

Thus he believed that this was not a states rights issue but the federal government should have the power to *"Prevent the increase of it."*

That raised a lot of eyebrows, since Mason was a slaveholder.

It infuriated Charles Pinckney of South Carolina who blurted out, *"How can slavery be wrong when it is legal all around the world? Throughout every generation in history at least one half of the people were slaves."*

"Today modern nations like France, England, and Holland still protect it. Leave the South alone and they will stop it on their own."

John Dickinson of Delaware was quick on his feet to point out that the slave trade should be abolished because *"Every principle of safety and honor will not allow it."*

Even England and France wouldn't allow it in their mainland but only in their colonies because they *"Recognized the danger of it."*

Elbridge Gerry of Massachusetts agreed, demanding that since slavery should be destroyed, they must put nothing in the Constitution that would sanction it. (He would later refuse to sign the Constitution because he felt it had failed to protect human rights.)

As the debate went on and on, it became clear that South Carolina and Georgia had drawn a line in the sand.

John Rutledge made everyone understand that they *"Would never be stupid enough to give up their rights."*[134]

James Madison spoke up that he *"Thought it wrong to admit in the Constitution the idea that there could be property in man."*[135]

His words would be ignored. As James Madison would later write to Thomas Jefferson,

"South Carolina and Georgia were inflexible on slavery. The result is seen in the Constitution."[136]

The end result was a Constitution that contradicted itself. It claimed that no one could be deprived of life, liberty, or property without due process of law. Yet it blocked any law against slavery from being passed for twenty years.

Slavery would be prohibited in the new western territories but escaping slaves could be pursued and recaptured in any territory.

Once again, battle lines had been drawn between freedom and oppression. America had split down the middle, even while it was in the process of uniting under a new federal government.

The compromise made to save the union, would put the union on a path to being ripped apart. The evil would not be extinguished until hundreds of thousands of men had sacrificed their lives on the bloody battlefields of the Civil War.

Benjamin Franklin protested this version in a final speech to the Convention. Voicing his disapproval very loudly he said he felt forced to accept it because he *"expected no better"* from that group.[137]

There were just too many cold hearts unwilling to do the right thing.

Before the Constitution took effect, it had to be approved by each state. For the next several months, each individual state held debates on whether to accept the Constitution. Once again, the topic of abolishing slavery came up in these debates.

In Massachusetts, many of the delegates were deeply grieved that the Constitution had failed to abolish slavery.

James Neal refused to sign the Constitution for that very reason.

General Samuel Thompson was furious over how Washington had failed to use his powerful influence to abolish slavery. *"Now that we've established our own freedom, how can we make slaves of others?"*

"O Washington! What a reputation he had! But he has proven he only cares for himself, because he holds those in slavery who have just as much right to freedom as he does."

Representative Isaac Backus, who was also a pastor, stood up next. He ripped apart all the proslavery arguments by showing from the Bible how much God hated it.

Then he concluded that while the Constitution was far from perfect, *"At least now a door has been opened to abolish slavery."*[138]

Surprisingly, thousands of miles away that's the same thing that Patrick Henry was telling the Virginia legislature.

Patrick Henry was against a strong federal government, fearing that it would eventually interfere with the rights of states. He made it very clear that if they approved the Constitution then it was only a matter of time until the federal government freed all the slaves. Despite his fiery speech, Virginia ratified it.

Robert Carter had wanted to attend those sessions on ratifying the Constitution in Virginia but things hadn't worked out. Like the rest of the nation, Robert Carter watched news of these proceedings closely, knowing that the outcome would determine the future of the nation.

He had even hoped that the Founding Fathers would figure out a plan for abolishing slavery. Then when news came the new Constitution had made a compromise with evil, Carter was disturbed.

He wrote to a friend that he was worried about the future of America, knowing God would hold the nation accountable for allowing the evil to continue.

He wasn't the only person deeply troubled by this.

As soon as the very first Congress under the new Constitution met in 1790, it received several petitions from Quakers pleading for the destruction of slavery. Many of these petitions reminded Congress that since America was a Christian nation, it had a duty to obey Jesus' command to *"Do unto others as you would have them do unto you."*

That launched a very heated debate. Many Congressmen were offended that anyone would preach to them. As the debate escalated, James Madison of Virginia stood up. He told Congress to think about how, *"If there is anything within the Federal authority to restrain such violation of the rights of mankind it should be done."*[139]

His motion was opposed by James Jackson of Georgia, who was deeply offended that the Quakers thought they had better morals than anyone else. He pointed out that Quakers had refused to serve in the military during the war.

While he recognized it was *"The fashion of the day to favor the liberty of the slaves,"* he disagreed with this philosophy.

"Are the Quakers the only ones whose feelings matter? Did they write the Constitution or help us win the war?"

"Do they know more about God's will than we do? If they would bother reading their Bibles, they would realize that slavery is God's command."[140]

After a lengthy debate, Congress decided to table the petition. That was the nice way of saying nothing else would be done.

Just when they thought the debate was over, the next day Congress received a special petition from Benjamin Franklin as President of the Pennsylvania Antislavery Society.

This was a petition that could not be ignored. It could not be swept under the table because everyone knew Benjamin Franklin had done something that no one else did.

He was the only person to have signed all four major documents of the American Revolution:

- The Declaration of Independence
- The Treaty of Alliance with France that turned the tide of the war
- The Treaty with England that ended the Revolutionary War
- The Constitution

Yet there was another paper Franklin had signed that was also really important. This petition demanded Congress recognize slavery could not legally exist in America.

After all the years he had spent helping America gain freedom, now Franklin was deeply concerned that America had forgotten the duty to *"Administer freedom without distinction of color."*

Quoting from the Constitution, the petition reminded Congress they had a duty to *"Secure the blessings of liberty to the people of the United States."*

Therefore since civil rights came from God and *"Equal liberty is still the birthright of all men,"* Franklin demanded that Congress utilize its full power to abolish slavery.[141]

There was no excuse for not dealing with the elephant in the room.

As soon as Franklin's petition was read, Congressman Smith of South Carolina was on his feet objecting. He reminded everyone that the South, *"Would have never entered the Union unless their property had been guaranteed. Besides my constituents don't need to learn their morality from this petition."*

That infuriated Congressman Scott of Pennsylvania who stood up and gave a long speech about how even if *"There were neither God nor devil, I should oppose slavery on the principles of humanity."*

Mr. Tucker of South Carolina disagreed. Disturbed about how other people were trying to force their morality on his state, he made it very clear that the South already had plenty of religious leaders to teach them whatever they needed to know. Besides, this petition requested something unconstitutional and Mr. Tucker was surprised to see that it was *"Signed by a man who ought to have known the Constitution."*

Elbridge Gerry of Massachusetts, who had protested slavery at the Constitutional Convention, protested it again by saying Congress had *"just as much of a right to regulate this business as we have any rights whatsoever."*

The debate went on for several days. Finally it was proposed that a committee be organized to respond to Franklin's petition.

One week later, the committee reported back that after researching all the relevant laws it had been decided, *"Congress has no authority to interfere."*[142]

That was supposed to end all debate on slavery. Yet it only caused the debate to grow louder, as Congressmen took sides. Again America split right down the middle on this issue, while the Civil War was still decades away.

The debate in Congress grew so intense that finally James Madison stood up and reminded everyone that under the new Constitution, Congress could not pass any laws regarding slavery for the first twenty years of the nation's existence.

Then Madison mentioned that while the committee had been reviewing the petition, Benjamin Franklin had passed away.

Madison proposed they all honor his memory by wearing a badge to recognize his service.

The subject was laid to rest.

Yet the American people would not stop. Antislavery petitions continued pouring into Congress until Mr. Rutledge of South Carolina objected that all these petitions *"Were continually coming in and stirring up strife. There's too much of this philosophy of liberty and equality."*

On the other side, Congressman Bard of Pennsylvania grieved, *"Our hands are tied. We are forced to stand confounded while we see the flood gate opened, pouring incalculable miseries into our country."*[143]

All this debate was published in the newspapers.

Carter read this and was disturbed the abolition movement was encountering so many roadblocks.

Writing to a friend, Carter worried about the future of his nation. He grieved over *"Much stupidity and hardness of heart"* in the leadership.[144] It was becoming clear that if something was going to happen he would have to do it.

CARTER DOES THE IMPOSSIBLE

For many years, the Holy Spirit had been dealing with his heart that slavery was wrong. That it wasn't enough to say it was wrong but do nothing like Patrick Henry and Thomas Jefferson had.

The Holy Spirit kept tugging at his heart. If he really loved God, he was going to have to prove it by freeing all of his slaves. Yet how could he do that without the legal system overturning it?

Carter was well aware of the long legal battle that his neighbor, Robert Pleasants, had fought to enforce the terms of his father's will to free the slaves. The will had been strongly disputed by family members. Carter realized he would have to figure out an ironclad plan. After consulting with lawyers, he developed a very complicated plan of *"Immediate but gradual emancipation."*[145]

On September 5, 1791, he filed legal paperwork at the courthouse notifying everyone that he intended to begin freeing more people than anyone else ever had.

Starting on January 1, 1792, on the first day of every year, groups of slaves would be freed, moving in order from the oldest to the youngest. He also made arrangements for the slaves to continue living in their homes on his land.

He planned a specific schedule because freeing them was sacrificing an estate which could have been sold for $100,000 in 1791. (Worth about $2,500,000,000 in today's money.)[146]

Carter worried that his children would bitterly resent losing their inheritance and challenge it in court. That could lead to the court seizing his assets and returning everyone to bondage. So he spent considerable time and effort on legal advice, figuring the more complicated he made the process, the less likely a court could rule it had been a mistake.

Then Carter wrote out a legal document called *The Deed of Gift* to make his intentions very clear. He wrote, *"I have for some time past been convinced that to retain them in slavery is contrary to the true principles of religion and justice. Therefore it was my duty to free them."*[147]

The *Deed of Gift* split his family right down the middle. Some family members helped him, including his daughter Julia, who carefully followed his instructions and was still freeing the slaves, years later in 1852.[148]

Others tried to hinder the plan, including his son John Carter, who vowed to *"Do all in his power to frustrate the scheme of emancipation."*[149]

Carter had prepared for it by leaving a large inheritance of land to his children that they would only receive, if they made sure the slaves were freed. Then because Carter still didn't trust his family, he put a nonfamily member, Benjamin Dawson, in charge of the estate because Dawson would make sure the plan was followed.

Dawson was a terrible money manager who made mistake after mistake in handling Carter's finances. Yet the one thing he did right was freeing people according to schedule. This was a very expensive process that involved significant court fees and paperwork but Carter had provided funds to pay all of those fees. Dawson stayed on schedule, freeing each group as had been planned.

Carter's family hated Dawson. Even though Carter had left more than enough inheritance to take care of his children for the rest of their lives, the backlash to his plan would be overwhelming.

Family members filed lawsuits, trying to overturn the will. One of Carter's sons tried to beat Dawson to a pulp. Another son would fight through the court system for years, trying to stop the plan. Yet all the years of Carter's careful planning would pay off. The Virginia courts would uphold his legal document.

Carter's neighbors were also furious at him. Many of the people renting land from Carter, demanded that their own rent should be lowered, since Carter must not need any money if he could afford to free his slaves.

In the years that followed, Carter's income dropped dramatically, as he encountered complaint after complaint from his tenants. He lost so much money that he ended up very deep in debt.

Then there was another problem. Carter had accidentally freed the neighbor's slaves as well. Apparently, many other neighboring slaves were able to escape by forging freedom papers, which looked just like the papers that had been given to Carter's people.

Neighbors complained that it was as if Carter had set fire to his own house and then burned down the entire town because of the frustrating *"Consequences of the Deed of Gift."* They pointed out there was *"Vastly greater number of these people now passing as your free men than you ever owned."*[150]

This resulted in Virginia passing a law making it illegal to forge freedom papers. Another law was passed requiring freed slaves to obtain new freedom papers every three years, which cost money they didn't have. Yet no matter how many obstacles came, those freed slaves would not allow anyone to take their new lives.

Many of Carter's friends turned against him. He was ostracized from polite society. Eventually the backlash grew so strong that he was forced to leave Virginia. He moved to Baltimore, Maryland. Just a few years later, he passed away in 1804, at the age of seventy-six. He hoped his family would understand: *"My plans and advice have never been pleasing to the world."*[151]

As soon as he was gone, his son George Carter, filed another lawsuit challenging the will and attempting to *"Take immediate possession of the remainder of the Negroes."*[152] The freedom schedule came to a screeching halt as this lawsuit went through the system. It took a few years but eventually Carter's thoughtful planning would work.

The court upheld the *Deed of Gift*. The remaining slaves were freed, including the ones in George Carter's household. Year after year, Dawson continued following the plan. When Dawson passed away, other trustees continued freeing people all the way up until the Civil War.

The *Deed of Gift* enabled lots of people to start a new life. Yet many of the historical details would be lost, because this story was suppressed, lest anyone be inspired to follow his example.

While lots of books would be written about the other founders of America, Carter's name was wiped from Virginia's history. He was buried in an unmarked grave as his legacy was forgotten.

Carter had never wanted any public recognition. He had only wanted to please God's heart. He would pass on to his reward in Heaven as his work continued helping others.

His freed slaves developed new communities across Virginia, many of which would later play an important role in the development of the Underground Railroad. The descendants of these slaves made their own mark on history. Some served in the Union Army during the Civil War, like Nimrod Burke who became a First Sergeant. Others, like Isaac Gray, became successful entrepreneurs. By 1860, Gray had built his own estate of almost seventy acres of land. [153]

The rest of the details are lost to history but one thing is clear. Carter was willing to upset his family, lose his friends and sacrifice a fortune to obey God.

In the years that followed, his neighbors were thinking about what he had done.

Just a few years after the *Deed of Gift* made national headlines, George Washington was thinking about his future. How would God judge his life in the courtroom of eternity? He told his close friends how he had come to regret being a slaveholder. That he was wondering if what he had done *"Would not be displeasing to the justice of the Creator."*[154]

Then he would surprise his community by leaving a will that freed his slaves. While no one really knows what inspired Washington to do that, it might have been his friend and neighbor, Robert Carter.

8

Historical Background
Part Four

What would happen if a maximum security prison decided to open its doors and turn loose all of its worst inmates on the nearest street corner?

Would violent crime rise? That actually happened in the earliest days of America.

In 1492, when Spain gave Christopher Columbus the finances to discover America, no one really wanted to go with him. People believed that the world was flat and they could fall off it if they sailed too far. His new idea that the world was round and you could sail east by sailing west seemed very unlikely.

So when Columbus couldn't get enough of a crew together, Spain recruited sailors by offering prison inmates the chance at a new life. Several volunteered and traveled with Columbus on multiple voyages.

Many years later, when England began to explore America, the idea was proposed to eliminate overcrowding in British prisons by sending the convicts to America.

At the time, England had been struggling with starting any new colony in America. The earliest expeditions by Sir Walter Raleigh had all ended unfavorably.

The first group of settlers gave up and returned to England. The next group died too quickly. The third group completely disappeared. British authorities wondered how could they establish a permanent settlement if no one else wanted to go?

In 1597, England passed a law that anyone considered to be an idle vagrant could be shipped overseas against their will. Of course, that meant law enforcement was free to define vagrant as anyone that they didn't like. Poor people were seen as a disposable commodity. No one cared that the poor people might have civil rights.

In 1606, Sir John Popham approached the British Attorney General with a plan for forming the Virginia colony. The proposal was to take the undesirable people of society: the poor, homeless, convicted felons and *"Idle vagrants that will not work, whose increase threatens the state"* and ship them off to America. His proposal was approved by King James I.

Sir John Popham quickly got to work organizing a group of convicts for this first settlement. According to John Aubrey, a historian of that time, Popham literally, *"Stocked Virginia out of all the jails of England."*

Things didn't work out as planned. The first group of settlers never made it to America. On the way there, their ship was captured by the Spanish. The passengers on board were sold into slavery and never heard from again.

Popham tried again, putting together a new group of settlers which included regular people who wanted to try the adventure of moving to the new world. As you remember, this was the group that didn't want to work, so they ran out of food and almost starved.

When news of their struggle for survival reached England, the Virginia Company decided to find a new supply of cheap labor for the colony. At the time, the city of London was very concerned about rising numbers of poor people and street children. The mayor of London worked out a deal with the Virginia Company. Young people would be kidnapped off the streets and taken to America to work without pay on the plantations.

This happened during the summer of 1618. Constables were ordered by the Lord Mayor to *"Walk the streets and apprehend all vagrant children, both boys and girls."*[155]

The city of London even paid £5 per child to cover transportation costs. That subsidy would lead to kidnapping becoming one of the most profitable industries in England. Over the next hundred years, British merchants made large sums of money by stealing thousands of children off the streets and sending them to America.

This massive human rights violation was encouraged by King James who saw it as a way of supplying cheap labor to the colonies. He even directly participated in it.

When local teenagers made too much noise and disturbed his hunting expedition, he ordered them to be shipped to America and sold into slavery, even though it was a direct violation of British common law dating back to the *Magna Carta.*

The families of kidnapping victims fought back. They filed lawsuits against the merchants but few of them would be successful because law enforcement protected the kidnappers.

When Elizabeth Brickleband disappeared, her family filed a lawsuit against John and Jane Dennison. The court found them guilty but only gave them a slight punishment. The Dennisons were sentenced to a few months in jail and a small fine. Court records showed they had been getting paid £9 per person that they forcibly transported to America.

The most well known kidnapping lawsuit of that time was James Annesley. As the only heir to one of the largest family fortunes in England, Annesley was envied by a lot of people. Even his own uncle wanted him out of the way so he could steal the family fortune.

Annesley was kidnapped by his uncle and sold into slavery in America. He worked without pay for over ten years on a Maryland plantation before finding a way to escape. Eventually he made it back home to England. Then he filed a lawsuit to get his own property back from his uncle. It took a while but he was eventually able to prove himself as the rightful heir and regain his fortune.

During this time, there were also many voluntary indentured servants as well. However, these quickly realized that they were at the mercy of the planters.

Thomas Hellier was an Englishman who signed a contract to go to America for a teaching job. He was promised that no manual labor would be required. Soon as he reached America, he was sent out to the fields and forced to work with his hands for extremely long hours. When he complained, he was brutally beaten.

He ran away several times but was caught and punished severely. Eventually, the constant abuse would cause him to take matters into his own hands.

Late one night, he killed the abusive planter and his wife. He was arrested, tried and convicted of murder. At his execution he gave a long speech about how the indentured system was violating the rights of the people. His voice would not be heard. Too much money was being made from the evil system that the planters would not allow anything to get in their way.

Meanwhile, England continued using America as a dumping ground for convicted felons. In 1718, England passed a law formally organizing the system of convict transportation. They were sold as indentured servants to save on the costs of incarcerating them. Terms of service ranged from seven to fourteen years, depending on the severity of the crime for which they had been convicted.

Buying these convicted felons was much cheaper for the plantation owners than buying regular indentured servants because the plantation owners wouldn't have to pay the convicts anything at the end of their term of service. (Regular indentured servants received land, tools and seed to start their own farms upon completion of their term.) To sweeten this deal, England even offered more subsidies to merchants willing to transport them to America.

Of course, things didn't work out according to plan. Several ships transporting convicts experienced violent mutinies at sea.

In 1718, thirty convicts being transported, broke free of their chains, seized control of the ship, sailed to France and disappeared.

In 1735, forty convicts from Ireland also took control of the ship, sailed to Canada, killed the crew and vanished. That didn't slow down the steady stream of convicts being transported.

Over fifty thousand were shipped to America in a time when the entire population of Boston was only fifteen thousand.[156]

Many of the convicts, who actually did make it to America, ran away as soon as they were sold to the plantations. Some tried to serve their full sentence but were treated so brutally that they died on the plantation before their time was up.

Others survived the years of hard labor, were released and became productive citizens.

Jonathan Ady was convicted of petty theft in England at a time when thieves received the death penalty. Ady begged for his sentence to be reduced to indentured servitude. His wish was granted. He was shipped to America where he would complete the indentured term, then start a new life.

He became a very successful farmer, served in the American military during the Revolutionary War and then came home to raise a family. He lived to the age of eighty-two, leaving a generous inheritance to his eleven children.

There were many others who also managed to beat the evil system of indentured servitude.

Matthew Lyon was an Irish teenager whose father had died fighting for land rights in Ireland. At fourteen years old, Lyon was punished for his father's work in fighting the hostile takeover of private property in Ireland. Lyon was sold into white slavery, transported to America and purchased by a Connecticut businessman, who resold him for a profit.

After ten years of hard labor, his way out came in 1775 with the Revolutionary War. Lyon ran away, volunteered for the military and proved himself under Commander Ethan Allen and the Green Mountain Boys, rising to the rank of Colonel in the military.

After the war, Lyon was elected to Congress. He fought hard for the rights of the common people, even getting into a fistfight on the floor of Congress when he was attacked by another representative.

He also became the only Congressman in American history to win re-election while locked up in jail. Upon his return to Congress, he managed to overturn the legislation that had imprisoned him by making it a crime to question the federal government.

Many of the convicts transported to America were actually good people who had resorted to desperate measures to feed their family in a time when England had a very unfair law enforcement system. However, there were also thousands of hardened criminals shipped to America during this time in history. They caused all kinds of problems.

As the convict population increased in America, violent crime also rose. Organized gangs carried out well planned robberies, terrifying families. The American people protested that their communities were being treated as dumping grounds. They tried to stop the inflow.

In 1722, Virginia passed a law closing its ports to convict ships. This law was quickly overruled by England, when the British merchant holding the contract for convict transportation, petitioned the Board of Trade. Maryland also tried to pass a law to protect itself from convict transportation. That law was also overturned by British authorities.

Benjamin Franklin complained about this in his newspaper.

Apparently several convicts had formed a street gang that was breaking into local homes and businesses. People were getting mugged by them on the streets in broad daylight.

Franklin pointed out that since England would not allow America to make any law that interfered with England's ability to profit from its colonies, then maybe America should respond by sending a shipload of rattlesnakes to England for every shipload of convicts that arrived here. At least rattlesnakes were polite enough to warn people before they struck.[157]

The only thing that stopped England from sending its convicts to America was the Revolutionary War. That closed all American ports.

British authorities believed that this was only a temporary problem. America would soon lose the war and the ports would open back up to convict transportation.

So during the war, England housed its convicted felons on floating prison ships. That way they would be ready for shipment to America whenever the ports reopened. To their surprise, things would turn out much differently.

When America won the Revolutionary War, King George III was upset. He tried to punish America by resuming convict transportation.

He approved the plan for a merchant named George Moore to transport 143 inmates to America onboard the *Swift*.

Moore was paid £500 and allowed to keep whatever proceeds could be made by selling these criminals in America.[158]

Once again, things didn't turn out as planned. The criminals seized control of the ship, turned it back to England and attempted an escape. Several criminals escaped from the ship only to be captured on land.

The remaining one hundred convicts, allowed themselves to be transported to America, then disappeared as soon as the ship docked in America, causing the merchants to lose a lot of money.

Moore tried again, sending another ship with 179 convicts to America in April of 1784.[159]

No port in America would accept the ship. Moore ended up having to unload his cargo in the British colony of Honduras.

Back in America, Congress quickly passed a law forbidding any convict ships from entering American waters. Once the profits dried up, convict transportation to America ended.

Parliament complained that now they had to figure out a new plan since *"The ports of America were closed against the importation of convicts."*[160]

The solution was to transport the convicts to Australia. Parliament complained at this new inconvenience since the cost of shipping them to Australia was so much more expensive.

Convict transportation was one of the many reasons that America declared independence. The main reason was Americans believed in the idea that everyone was created equal and governments existed for the protection of human rights.

For thousands of years, most of the world had believed that the common people only existed to serve their lords and masters.

Generation after generation, tyrants had ruled their people with an iron fist. It was only natural that England would rule America the way that most European nations ruled their colonies, believing that the purpose of America was to benefit the British Empire.

England tightly controlled America's economy by regulating all imports and exports. Merchants in America could only sell their products to England so that England could sell them to the rest of the world at higher prices. America tried to negotiate better trading partnerships but England wouldn't listen. Instead, England found ways to profit from unfair monopolies such as requiring America to only buy tea from British companies.

In 1766, Parliament passed the *Declaratory Act,* proclaiming they could pass any law on America that they wanted. America had no right to interfere.

America fought back.

In the city of Boston, Paul Revere, John Hancock, Samuel Adams and John Adams (they were cousins) formed the secret organization *Sons of Liberty* to organize protests.

They got the attention of the British when they snuck on board ships in the Boston harbor and dumped all British tea overboard. Since this was tea that the colonists were required to buy, England retaliated by shutting down the port of Boston.

Then England stripped Massachusetts of its local government by passing the *Intolerable Acts,* revoking all government officials who had been elected by the people. Going forward these positions would only be filled by men appointed by the King. Courts in America were also placed under British control. More British soldiers were sent to keep the peace in America and the Americans would have to pay for their room and board.

This was exactly what Benjamin Franklin had been warning them about. For years he had been telling the American colonies that they needed to start working together.

Seeing this storm coming, he had published many newspaper editorials warning how England was going to tighten its control of America. When no one would listen to his warning, he found something that they would hear.

In 1773, Franklin published private letters that had been intercepted between the British Governor and Lieutenant Governor of Massachusetts, describing how they were going to make harsher regulations for Boston. These private letters showed a conspiracy to crush the colonies. Things were going to get worse.

Franklin warned the leaders of America,

"If we don't all hang together, then we will definitely hang separately."[161]

As things went from bad to worse, America had to figure out what to do. At that time each colony had a local legislature, which had been elected by the people. These legislatures met together and decided to break the law.

Each legislature selected a few delegates and sent them to Philadelphia, Pennsylvania to hold the first ever meeting of the Continental Congress. Virginia sent George Washington and Patrick Henry.

Massachusetts sent John Hancock, Samuel Adams and John Adams. The New York delegation included abolitionist John Jay who would later serve on the Supreme Court.

For the first Continental Congress, fifty-six delegates met together in September of 1774. While the purpose of this meeting was to respond to the *Intolerable Acts,* many of these men would become the Founding Fathers of America. First they had to figure out what to do. Would they keep trying to negotiate better terms with the King of England? Or would they declare independence? Trapped between two very difficult choices, there was no easy answer.

While they had been suffering under England's policies, they didn't have enough resources to fight a war against someone with the most powerful military of that time.

As the Continental Congress later admitted to the nation, when they called for a day of fasting and prayer, they knew they were defying, *"A gigantic adversary."*

They were trying to do it *"Without arms, ammunition, revenue, government or ally. With only a staff and a sling (like David faced Goliath) we dare only to boast in the strength of the Lord of Hosts."162*

What happened next would be described by John Adams in a letter to his wife, Abigail Adams, dated September 16, 1774.

"When the Congress first met, Mr. Cushing made a motion that it should be opened with prayer. It was opposed by Mr. Jay of New York and Mr. Rutledge of South Carolina, because we were so divided in religious preferences."

"It was thought that since some of us were Episcopalians, some Quakers, some Anabaptists, some Presbyterians, some Congregationalists that we could not join in the same act of worship.

"Samuel Adams stood up and said that he was no bigot and could hear a prayer from a man who was a friend to this country. Therefore he moved that Mr. Duché, a local Episcopal clergyman, whom he had heard had a good reputation, might pray the following morning."

"The motion was seconded and passed. Mr. Randolph (the President of the Congress) visited Mr. Duché and invited him to come. The next morning he came and read Psalm 35."

Psalm 35 described how powerless the leaders of America felt at that time. In it, King David cried out to God for help, saying:

"O Lord, Fight against those that fight against me."

"Let not my enemy rejoice over me."

"The Lord delivers the poor from those who are too strong for them."

"My soul shall rejoice in God's salvation."

Adams described to his wife how this Psalm greatly encouraged them. *"I've never seen a greater effect upon an audience. Remember this was the next morning after we heard the horrible rumor of the cannonade of Boston."*

"It seemed as if Heaven had ordained that Psalm be read on that morning. Then Mr. Duché ended with a fervent prayer, which filled the heart of every man present. He prayed for America, for Congress and especially for Boston. This had a considerable effect on everyone here."[163]

During that prayer, George Washington had been on his knees.

Mr. Duché would be appointed as the first ever Congressional Chaplain.

Congress would issue several proclamations for national days of prayer and fasting for divine intervention.

Ever since, opening government meetings with prayer would be an American tradition. Yet first America had to become a nation.

First prayer in Continental Congress

The First Continental Congress would decide to try one more time to negotiate with England. If that didn't work then they would meet again to form a plan. In the meantime, they organized a boycott of all English imports until England changed its repressive policies.

They also planned a second meeting for May 10, 1775 to deal with whatever happened next.

What happened next was the American Revolution.

British authorities heard the Americans were conspiring in secret meetings. Several hundred British soldiers were sent to Boson to capture Samuel Adams and John Hancock because they had been the most publicly visible in stirring up the people towards revolution.

Paul Revere's Midnight Ride

Paul Revere was waiting for them. His midnight ride woke up the Americans, enabling Samuel Adams and John Hancock to escape.

When the British soldiers arrived in the Boston area, everyone was waiting for them. Word on the street was that the British had been ordered to seize all weapons and ammunition. That was true but that wouldn't happen.

John Ward Dunsmore painting of Marinus Willett preventing the removal of arms by the British on June 6, 1775.

The minutemen of Boston stood their ground, standing in the way of the British troops. According to Paul Revere, who watched it all unfold, when the British arrived at Lexington, Major John Pitcairn yelled, *"Lay down your arms, you damned rebels!"*

American Capt. John Parker issued orders to his men, *"Don't fire unless fired upon. But if they want to have a war let it begin here."*[164]

The British fired the shot heard around the world. The war began and spread very quickly. By the time that the Continental Congress met again on May 10, 1775, the war was in full motion.

Congress sent George Washington to lead the military and fight the war.

Washington was ready. He had prior military experience and had worn his military uniform to every session of the Continental Congress.

Now it was time for him to earn that uniform by serving in combat. Washington hurried to the front lines of the conflict and succeeded in forcing the British to withdraw from Boston.

Meanwhile, Congress tried one last time to negotiate peace with England.

Just like all the other times they had tried to negotiate, this one wouldn't work either. Instead they found themselves trying to plan a war and lead a new nation.

They knew that they had to explain to the world why these colonies could declare independence as a sovereign nation. It wasn't just about taxes. It was this crazy idea that everyone had been created equal. That people had civil rights and that everyone was accountable for their actions.

Congress selected a team to write the *Declaration of Independence.* On that team was Thomas Jefferson, Benjamin Franklin and John Adams.

Thomas Jefferson wanted John Adams to write it but Adams knew Jefferson was the better writer and convinced him to draft it. This would become one of the most important documents in the history of the world.

Together they wrong a long list of their grievances with the King:

- *"He has made judges dependent on his will alone for the amount and payment of their salaries."*
- *"He has created a multitude of government positions and sent swarms of officers to harass our people and eat their substance."*
- *"He has cut off our trade with the world."*
- *"Plundered our seas, ravaged our coasts, burnt our towns and destroyed the lives of our people."*

As they wrote the *Declaration*, the founders of America still didn't know if democracy could actually work. If they could actually defeat a world power. Knowing they didn't have enough resources to make this stand, they threw themselves on the mercies of God.

Written into the *Declaration* was an appeal *"To the Supreme Judge of the World."*

Independence was declared *"With a firm reliance on the protection of Divine Providence, we mutually pledge to each other our lives, our fortunes and our sacred honor."*[165]

While Thomas Jefferson wrote most of the *Declaration*, once it was finished, it would be John Adams who had the even harder job of convincing Congress to sign it. They did, even though they knew the road ahead was very difficult. The only path to freedom was through struggle and suffering.

Many of the delegates were worried about the future. Adams knew exactly how desperate their situation was.

On July 7, 1776, John Adams wrote to his wife how things looked very bleak for America. They had signed the *Declaration* while wondering how they were going to pay for it. How could they fight a war while the military was suffering so much that the situation had become *"horrifying?"*

The army had *"No clothes, beds, blankets or medicines and no food but salt pork and flour."* Morale was dropping as they were being overrun by disease. The soldiers were feeling *"Disgraced, defeated, discontented, dispirited, diseased and eaten up with vermin."*

American soldiers in the Revolutionary War

John Adams wrote, *"Yet through all the gloom I can see the rays of light and glory."*

"While I am well aware of the toil, blood and treasure it will cost us to maintain this Declaration and support and defend these states, I can see that the end is more than worth all the means. And that our posterity will triumph even though we might regret it, which I trust in God we won't."

When his wife read that letter, she wrote back, *"I feel no fear at the powerful weapons designed against us. The God who fed the Israelites in the wilderness and who clothes the lilies of the field will not forsake us. I pray daily for you that you may have health, wisdom, endurance and that your labors be successful."*[166]

The *Declaration of Independence* would send shock waves around the world. At that time, America was not the only colony of England. The British controlled an enormous empire that stretched around the world from the rising to the setting of the sun.

What would happen if all the other colonies decided to revolt? Would they suddenly realize that they had God-given rights?

The possibilities were frightening for some people and exhilarating for others. When this news reached France, it captured the heart of a young French nobleman named Marquis de Lafayette.

Portrait of Lafayette

Lafayette in France with King Louis XVI and Queen Marie
Antoinette

9

Marquis de Lafayette

"Greater love has no man than this,
that he lay down his life for his friends."
John 15:13 (BSB)

Marquis de Lafayette
1757-1834
From France

Nineteen year old Lafayette was the sole heir to one of the largest private fortunes in France. Just the annual rental income from his properties would be about one million dollars a year in today's money. While the money made his life very comfortable, Lafayette still had a heart for serving his country.

Lafayette's family had served in the French military all the way back to his ancestor, Gilbert de Lafayette III, who had fought side by side with Joan of Arc. Lafayette's own father had laid down his life for his country, fighting at the Battle of Minden.

Lafayette followed in their footsteps, serving in the French military. He studied advanced warfare tactics at a prestigious military academy in France, becoming one of the youngest commissioned officers.

When Lafayette wasn't fulfilling his military duties, he was moving in the highest social circles of Europe. He married a beautiful lady who was a direct relative of the King of France. They were often guests at the royal palace and invited to all the best parties. Lafayette was even a dancing partner of Queen Marie Antoinette.

When he visited England, he was given a seat of honor at the banquets of King George III. While Lafayette enjoyed the kind of life that most people could only dream about, he dreamed of something bigger.

When he heard of the *Declaration of Independence*, Lafayette was inspired. Maybe this was the beginning of a worldwide revolution.

France was still suffering under the ancient system of feudalism. While Lafayette was at the top of that system, with the power of life and death over the peasants that worked his land, he believed in the idea that everyone had been born equal.

He hoped that if democracy could work in America then it could spread around the world. The old traditions and class distinctions would disappear and new opportunities would open to everyone. Knowing how difficult it was for America to challenge such a powerful adversity, they were going to need all the help they could get. He wanted to be involved.

In 1776, Lafayette heard that America had sent an ambassador to France to ask for help. Ambassador Silas Deane begged King Louis XVI for help but the King was reluctant to get involved.

France and England had been at war for generations. When America was discovered, both of them laid claim to the new land. France had colonized the Louisiana Territory and parts of New England until the French and Indian War (1754-1763) when England had pushed France back beyond the Mississippi River. That war had drained France's resources, leaving King Louis reluctant to get involved in another long-term conflict.

Deane promised that if France helped America, then France would have a powerful new ally and trading partner. The economy of France could greatly benefit from partnerships with American manufacturers.

Plus, America would be willing to help France if they were attacked by anyone else.

(Note: In World War I and World War II, America kept the promise to help France.)

While that offer was tempting, King Louis wasn't sure. He discussed this with his advisors in the French court. Word quickly spread to England that France was considering supporting America.

England threatened a full naval blockade of France's coastline if they interfered. Lafayette's own uncle, the French ambassador to England, convinced King Louis to avoid taking sides and instead issue an order forbidding any of the French people from helping America.

Lafayette disobeyed that direct order. He went to meet Silas Deane and offer his services, volunteering to pay his own expenses to go to America. Deane replied that a nineteen year old teenager was too young to be a commissioned officer in the American military. Lafayette replied he would be happy to enroll as a private. His strong humility impressed Deane so much that he wrote a letter of recommendation to Congress about this remarkable young nobleman.

While this was going on, the British Navy followed through on their threat to block French ports. That created a real problem for Silas Deane. By this time, he had been approached by French military officers willing to go to America, but there was no way of getting them there.

If they tried to run the blockade they might be captured by the British Navy and become prisoners of war. They could also be arrested by the French government, who had made it clear that anyone trying to help America would be punished. Just when Deane was about to give up and go back home, Lafayette came up with the solution.

LAFAYETTE FIGHTS FOR FREEDOM IN AMERICA

Lafayette volunteered to pay whatever it cost to charter a private ship willing to run the blockade and get to America. Any other military officers wishing to travel with him were welcome to come.

Then Lafayette actually had to sneak out of town. Since he was in between terms of active duty he was running out of time to leave before he would receive orders to return to serve in the French military.

He could not even say goodbye to his own family because they would have stopped him from leaving.

He wrote a letter to his family, telling them that he wanted to help America become a nation so it would help the whole world. Lafayette barely made it to the ship in time as the King issued an order directly forbidding Lafayette from leaving the country.

When Lafayette's family found out, they were furious. How could he leave his pregnant wife and young child? Plus, he could bring the King's wrath on their family. Yet that was the risk Lafayette was willing to take.

British ships on both sides of the Atlantic were sent to find his ship and arrest him, but Lafayette slipped out of their grasp and arrived on American soil, just a few months later. It was a miracle, he made it safely there. Just as his ship was approaching the coast of South Carolina, they had almost been captured by the British. Yet he escaped just in time.

Lafayette made it safely to America only to find out that there was a much bigger problem. When he finally reached George Washington's troops at Philadelphia, he was horrified to find out how much the American military was lacking supplies and equipment.

There was no money to fight the war because Congress didn't have power to raise taxes. The military was struggling to fight a war without equipment. Battles were already being lost for lack of ammunition.

As Lafayette looked around the military camp, he wondered how this raggedy bunch was going to defeat the well supplied British army.

As he later described to his wife he saw: *"About eleven thousand men, ill armed, and still worse clothed, presented a strange spectacle."*

"Their clothes are motley looking, discolored and many were almost naked. The best dressed wore hunting shirts, large gray linen shirts used in the Carolinas."[167]

George Washington himself was very embarrassed to meet him. What did he have to offer this young French nobleman, when he couldn't even take care of the soldiers he already had?

Washington wrote in his journal:

"Few people know the predicament we are in. To see men without clothes to cover their nakedness, without blankets to lie on, without shoes, marching until our movements could be traced by the blood from their feet and yet not complaining is remarkable patience and courage."[168]

As they sat down and talked, Washington apologized to Lafayette for the disappointment he must feel, seeing the desperate reality of the impossible odds that they faced.

Washington was surprised by Lafayette's response. He said that he had come to America *"To learn not to teach."*[169]

Lafayette vowed he would fight to his last breath for America. Washington was deeply touched by the humility and selflessness of this young man. He appointed him as an officer. It would be the beginning of a lifelong friendship between Washington and Lafayette.

But first there was a war to fight and things were looking worse by the day. The British General Lord Cornwallis was advancing on them with the full force of a large army. While Washington's troops were way outnumbered, they had to make a stand or Cornwallis would quickly overwhelm them.

Washington called a meeting of his top military generals to form a plan. As they debated what to do, Lafayette volunteered to lead the attack.

It was a very bloody battle. Lafayette tried to hold his ground but was quickly forced into retreat by a violent assault. Still he stayed on the front lines of the battle, boldly facing the danger, as he kept the troops as organized as possible.

The battle intensified. Many troops were badly injured. Lafayette was shot in the leg. As pain ripped through his body and blood poured out from his wound, he stayed on his horse and kept shouting orders to his men. He had to keep them moving or they would be quickly surrounded by the British.

Washington would later point out that even though America lost this battle and suffered heavy causalities, it was the courage and skill of Lafayette that saved the lives of many soldiers by moving them back in an organized manner.

When the battle was finally over, Lafayette had lost so much blood, he could barely walk. Back at the camp, Washington told the military doctors to care for him as if Lafayette was his own son.

News of Lafayette's courage and sacrifice quickly spread around the world. From the halls of Congress to the palaces of France, everyone talked about how this young nobleman had risked his life for America.

In a time when the nobility was known more for their extravagance than their sacrifice, the world was deeply moved by the courage of Lafayette.

As Lafayette began receiving letters of appreciation from home, he realized that he had just turned the tide of public opinion towards the cause of America. *"By quitting France in such a scandalous fashion I actually served the American Revolution!"*[170]

Washington used this to his advantage. While Lafayette donated large amounts of his own personal fortune to buy supplies, uniforms and equipment for the military, it was not enough to fight the war.

So Washington sent Lafayette to ask Congress for more military funding, hoping they would listen to him.

Congress greeted Lafayette with a hero's welcome but did nothing about his request.

There just wasn't any money to send. However, Congress was so impressed with him that later on during the war, at one point they fired George Washington and appointed Lafayette as Commander-in-Chief of the U.S. Military, answerable only to Congress.

Lafayette declined this honor, saying that he preferred being second in command to Washington to any other title.

Lafayette stayed on the front lines of the battle, working closely with Washington. He was put in charge of organizing an alliance with the Native Americans.

Washington knew there were Native American tribes who had been loyal to the French during the French and Indian War and would welcome someone like Lafayette. This was crucial for America because the British were also working behind the scenes, trying to persuade the Native Americans into joining forces with them.

Lafayette succeeded in building strong relationships with several Native American tribes, leading to many serving in the military. Some were scouts that kept track of enemy movements. Some were highly effective snipers that terrified the British. Others trained American soldiers on guerrilla warfare tactics.

Washington also put Lafayette in charge of what to do with all the other French military officers who had come to help. Most American soldiers were uncomfortable being put under the command of French officers, who did not speak English well and did things completely differently.

Lafayette created a unit of troops made entirely of foreign officers and soldiers. That way they could communicate in their own language and fight in the way they were more comfortable. Lafayette also served as a liaison between these French volunteers and the American officers, helping smooth over cultural misunderstandings.

When some of these officers became more of a problem than a solution, he solved that problem by paying their way back home.

Lafayette also had another strong advantage.

The time he had spent moving in the highest social circles of Europe had introduced him to many of the top ranking British generals.

He had listened to the generals at countless dinner parties in the palaces of France and England. Lafayette knew them well enough to predict how they would operate and defeat their tactics. This quickly made Lafayette one of the best assets in the war.

The British knew it too. They knew they had to capture Lafayette if they were going to win the war. So they sent a large group of British soldiers to hunt him down.

His position was betrayed by an American deserter.

Lafayette was warned in time to prepare but there was very little he could do. In a short period of time he was surrounded, outnumbered and heavily outgunned.

With the odds completely against him, Lafayette achieved one of the greatest victories in the war.

Once the British had completely cut off all his escape routes and avenues of supply, they were so confident of success that they actually planned a celebration party. Invitations were sent to British dignitaries telling them to come and see Lafayette in shackles. There was even a ship waiting to transport him back to England. But they would be outsmarted by a young man who knew that if he could frustrate or embarrass them enough, they would back off.

As the youngest officer in the war, Lafayette had a secret weapon. He always listened to his men. Even though he was one of the generals appointed by Washington to lead the soldiers, Lafayette would sit down with his soldiers and ask them for ideas and strategies. What had they learned from their combat experience? The way he listened to their suggestions earned him the respect of his men and gave him the strategy for this battle.

First, he made it seem like he had a lot more men than he really did. Lafayette told the drummers to spread out far apart to make it sound like they were issuing orders to a much larger group. Then he had small groups of soldiers advance on the British.

Hiding behind the trees, they fired at the British and then retreated quickly. Then another group advanced on the British from a different angle, fired quickly and disappeared.

British soldiers started dropping left and right, while no one could tell which direction the fire was coming from. It really frustrated everyone.

Lafayette described, *"In every direction the British looked, they saw nothing but the heads of false columns popping out among the trees and then disappearing."*

"At one point fifty savages, our (Iroquois) friends in war paint, suddenly came face to face with fifty English cavalrymen, who had never seen an Indian before. The war cries of each side so caught the other by surprise that they both fled with equal speed in opposite directions."[171]

While the British officers began arguing among themselves about what to do, Lafayette organized his men to retreat in a format that the British would follow. He had the soldiers move backwards, drawing the British after them, until they had caused two groups of British soldiers to come face to face with each other.

As he described, *"When the two British lines met they were on the point of attacking each other, for there was no longer anyone between them. General Howe was astonished. The whole British army, of which half had marched forty miles, retreated without a single captive."*

"The English returned to Philadelphia much fatigued and ashamed and were laughed at for their failure."[172]

This resulted in British authorities holding an investigation into why things had gone so bad. Meanwhile, Lafayette was moving his men out to where they could hold their ground.

As the British retreated, Washington saw the chance to attack them from behind, while they were frustrated and exhausted. Calling a meeting with his top military generals, Washington planned a surprise attack.

At the meeting, General Charles Lee tried to persuade everyone that the British military was too powerful for the Americans to risk this attack. Better to let them retreat while America worked on its future plans. Lee convinced everyone, except Lafayette.

When Washington asked the other generals for their opinion, Lafayette said it was time to press forward and fight.

Washington agreed but put Lee in command of the attack. That was a terrible mistake. Lee did nothing. Then he made excuses. Then he ordered the troops to move in the opposite direction. Lee led the troops into confusion and failure. While Washington was telling them to press forward, Lee was telling them to stay back. American soldiers ended up all over the place, advancing in some directions, while retreating in others.

Furious at this incompetent leadership, Washington took control. He led a dramatic charge forward on horseback, directing one group of soldiers, while Lafayette followed him with another group advancing from a different direction. They pushed the British out of Philadelphia and New Jersey, regaining some of the most important territory in the war.

It was a powerful victory. After the battle, Washington had General Lee court-martialed for *"Disobedience of orders in not attacking the enemy and misbehavior before the enemy in making an unnecessary, disorderly and shameful retreat."[173]*

It would later be revealed that General Lee had been a British spy, working against America.

That bothered Lafayette. He told Washington how when he first heard of America, he had thought that everyone there must be united in their love of freedom. How could they turn against America?

Lafayette was also by Washington's side when Benedict Arnold's treachery was discovered. They actually sat down for dinner with Arnold, at the base near West Point, not realizing Arnold had just sold West Point to the British.

His plot was that the British could cut off New England from the rest of America, by seizing control of West Point and blocking all movement up the river. Detailed intelligence of how to capture the fort had been sold by Arnold to the British for about one million dollars in today's money.

While Arnold's information was being smuggled to the British by a British officer named John Andre, Washington and Lafayette were eating dinner with Arnold at West Point, not realizing that British soldiers were on their way. The British knew that capturing both Washington and Lafayette could turn the tide of the war.

John Andre was captured by some sharp-eyed Americans, who recognized something about him didn't look right.

When news reached the American camp that treachery had been discovered, Benedict Arnold left immediately. He escaped before people realized who the traitor was.

Lafayette set off in hot pursuit of Arnold but was not able to find him. Lafayette returned to the camp and helped Washington pick up the pieces of an army scattered by treachery. This grieved his heart but motivated him to continue doing everything he could to strengthen America.

Lafayette spent the hardest winter of the war with Washington at Valley Forge. There was still a huge shortage of supplies.

Things looked bleak as Lafayette described in a letter to his wife. *"Our position was unsustainable. The soldiers lived in misery, lacking for clothes, hats, shirts, shoes. Their legs and feet black from frostbite. We often had to amputate."*

"As the army frequently went whole days without provisions, the patience of the men was a miracle. The sacred fire of liberty burned on."[174]

On December 23, 1777, Washington sent word to Congress that roughly, *"Three thousand men in the camp are unfit for duty because they are barefoot and otherwise naked. If nothing changes, this army will either starve, dissolve, or disperse to obtain subsistence."*[175]

What could Lafayette do? Already he had donated as much of his own personal fortune, as he could access while thousands of miles away from home. But it was going to take a lot more to win the war. Help was needed.

Thinking of his friends back home, Lafayette wrote a steady stream of letters to the leaders of France.

He appealed to their emotions, describing stories of remarkable bravery in the war and pleading with them to help America. He made Washington sound much larger than life, describing how powerful he was to watch in action and admiring his leadership.

The letters began having an effect. Benjamin Franklin and John Adams were in France at the time, trying to raise support. They saw Lafayette's words touching the hearts of people. They wrote back home that Lafayette's voice was being heard.

Everyone from the lowest to the highest levels of society thought of him as a national hero. After everything that England had done to France, in hundreds of years of war between the nations, the French people were thrilled to hear of the defeats of the British army. As Franklin and Adams continued lobbying France for support, the doors began to open.

One day, George Washington received a letter from Benjamin Franklin in France. This would be the letter that he had been hoping and praying for. With Lafayette by his side, Washington ripped open the letter and read the good news that France had signed a treaty recognizing America's independence from England.

This was a huge step towards winning the war. No longer was America just a colony struggling for survival. They had earned the respect of the world. Now France was promising to stand by them, even at the risk of war with England. While the British believed they could easily defeat the American colonies, they knew that they could not defeat France.

Supplies and equipment were still desperately needed. As George Washington ran out of options, he asked Lafayette for a big favor. Could he return to France and convince the French to send troops and equipment?

Lafayette didn't hesitate. He quickly booked passage on a ship back to Europe. Yet leaving the battlefield would put his life in even greater danger. The King of England had offered a big reward to anyone who could capture an American ship. There was an even bigger reward for the capture of Lafayette.

Lots of people knew about the reward, including the men working on the ship taking Lafayette back home. Some of them decided to mutiny, seize control of the ship and head for England to claim the reward.

Lafayette was aware of the reward on his head and had prepared for this type of scenario. Bringing a group of soldiers with him, he was able to fight back, crush the mutiny and retake control of the ship. He made it safely to France, where the mutineers were apprehended by law enforcement.

To his surprise, Lafayette received a hero's welcome in France. Massive crowds turned out to cheer his return. Lafayette used this new celebrity status to raise more support for America. He convinced France to loan America £6,000,000.

Benjamin Franklin was impressed. He had spent months in France begging for that loan, only to have his request denied over and over. Once again, Lafayette had opened doors that were closed to everyone else.

Lafayette's heart was still in America. As soon as he could, Lafayette returned to the front lines of the battle in America, just in time to face off against British Lord Cornwallis at Yorktown.

This was a very important battle. Lafayette knew that surrender was not an option. He had heard what had happened at the battle for Fort Griswold. That Fort had strategic value for controlling ships going up and down the Thamas River in Connecticut.

A small group of brave American soldiers had been guarding it. Then traitor Benedict Arnold led the British army to attack the fort. When the British came, the Americans stood their ground.

The battle was violent. The Americans fought back hard, killing many of the British officers. As long as their ammo lasted, the Americans held the British back. Once the Americans ran out of ammo, the British took the fort.

Trying to save the lives of his men, the American commander surrendered. He handed his bayonet to the British officer (not Arnold).

The British officer killed him with it.

The American soldiers watching were shocked.

Jordan Freeman and Lambo Latham were black American soldiers who had been defending the fort. Latham was standing near the British officer when the massacre started.

Latham's gun was out of ammo, but he still had his bayonet in his hands. He charged at the British officer, killing the man who had just killed his commander.

For that he would die a painful death. The British soldiers violently attacked him, stabbing him over thirty times. But his sacrifice was not in vain. He had successfully taken out a major British officer.

The British retaliated by slaughtering the other American soldiers, including Jordan Freeman. Then the British ransacked the town, taking what little food and supplies the women and children had left. To really hurt the civilians, they destroyed many buildings, leaving families without shelter.

That brutal massacre was on Lafayette's mind at the Battle of Yorktown. This was his chance to avenge the enemy's war crimes. He told the soldiers to remember Fort Griswold as they charged into battle.

Lafayette knew Lord Cornwallis well. They had partied together long ago in the palaces of France and England. Now they were fighting on opposite sides. But Lafayette had the advantage of knowing how Cornwallis would fight.

Lafayette used this to help America win by forcing Cornwallis into a defensive position, where his only option was surrender. After this battle, Parliament in England voted to end the war.

The miracle had finally happened. The struggling American colonies had defeated the powerful British Empire.

LAFAYETTE FIGHTS FOR FREEDOM IN FRANCE

Lafayette had accomplished his purpose. He returned home to France and to his life of royal parties and events. He continued serving as the national hero of France, with all the young noblemen of France dreaming of becoming like him. Even Queen Marie Antoinette found ways to honor him.

Yet as Lafayette returned to luxury, he felt something was missing. He stayed in touch with George Washington, writing to tell him, *"Ten times a day, I find myself wishing I was on the other side of the Atlantic. But our cause will be better served by my presence here."*[176]

As Lafayette moved back into the highest circles of French society, he kept his promise to Washington. Lafayette used every chance he could to lobby for the interests of America, trying to open doors for America to trade with the world.

People noticed that Lafayette showed up at every formal event, wearing the uniform of an American military general, making a powerful statement of where his heart was.

As the young American nation began trying to establish diplomatic relations with the world, some doors opened and some doors closed.

When America tried to send an ambassador to Spain and Spain rejected him, Lafayette got involved.

He knew the real reason Spain was doing this. Lafayette wrote to American diplomat Livingston that Spain, *"Feared the loss of their colonies. The success of our Revolution encourages this fear."*[177]

Taking matters into his own hands, Lafayette went to Spain.

Defying the rules of social etiquette, Lafayette boldly wore his American military uniform at the court of the King of Spain.

He showed Spain how it would greatly help them to work on a trading relationship with America.

By the time Lafayette was done negotiating, Spain had changed its policy. The U.S. Ambassador was received and ever since has been welcomed. This was the beginning of a warm relationship between the nations, opening the door to America later purchasing the Florida territory from Spain.

Lafayette returned to a quiet life, but things in France were far from calm. The people were starving. Desperation was growing from the reality of very serious food shortages.

Lafayette had food. His land had produced bumper harvests, which were stored away. When Lafayette saw the people's desperate needs, he moved quickly, ordering his property managers to open the food stores and feed them.

The managers told Lafayette that he could make a lot more money by selling the grain. Food prices had risen so rapidly, the profit margin was high. Lafayette refused. Instead, he launched a full investigation into why food prices were so high. What he found was very disturbing.

France still suffered under the ancient system of feudalism. The French economy was being controlled by a powerful monopoly known as the *Ferme*. This monopoly regulated the flow of all goods in France, for the purpose of imposing high taxes on them.

Revenue from these taxes financed the lavish lifestyles of the royal family and nobility of France, while making the common people suffer. Lafayette was horrified to realize the *Ferme* was intentionally driving up food prices by reducing food supplies.

Lafayette declared war on the *Ferme*. He made a public announcement that he would begin flooding the market with as much grain, as he could get from his lands. By increasing the supply of grain for sale and selling it outside of the monopoly, he hoped to break the monopoly and drive the prices back down.

Lafayette had a powerful effect on the French economy. He forced the *Ferme* to end the artificial food shortages, saving many lives and making him an even bigger hero to the nation.

He wasn't finished. Escalating his attack on the *Ferme*, Lafayette wrote a report exposing how it was destroying the economy.

Showing how France could prosper by granting free flow of goods instead of trade restrictions and monopolies, he wrote, *"The Ferme can do nothing but impede trade. Already France has lost the tobacco trade."*

"One particular tobacco ship waited in France for nine months during which the Ferme would neither buy the cargo nor permit it to go to Marseilles where the Italians wanted to buy it. These abuses are restricting our trade, hurting our citizens and offsetting the advantages we have. They also lead to smuggling and cheating."[178]

Helping him was Thomas Jefferson, who was living in Paris at the time as the U.S. Ambassador to France. Jefferson had also recognized the problems of the *Ferme*, writing how it was *"Contrary to the spirit of trade to carry a product to a market where only one person is allowed to buy it and of course that person fixes the price."*[179]

Jefferson suggested limiting the *Ferme* to only a 5% markup, so France's economy could grow. The increase in trading would still provide enough tax revenue to fund the government. France ignored Jefferson's suggestions but listened to Lafayette. Some trade restrictions were relaxed and others continued.

Lafayette pressed forward, lobbying French authorities on behalf of American businesses, hoping to open doors for more American goods to be sold in France. He was remarkably effective.

Thomas Jefferson wrote home how impressed he was by Lafayette. *"He is the most powerful advocate. While I'm only allowed to lobby at certain designated times, Lafayette has every door open to him, at any time, with great influence and connections. The truth is I only hold the nail. He drives it in."*[180]

There was another cause close to Lafayette's heart. Inspired by the success of the American Revolution, Lafayette dreamed of abolishing slavery throughout the world. He began corresponding with

Wilberforce's team of British abolitionists. They sent Thomas Clarkson to France, to help him lobby the leaders of the nation.

Lafayette invited Clarkson to stay at his home and opened doors to him. When Clarkson received death threats, Lafayette protected him, making it known that anyone who tried to hurt Clarkson would have to answer to him. They encountered intense opposition but Lafayette still used his time at the French court to personally lobby the King to abolish slavery. However, the King had other matters on his mind and declined to intervene.

That made Lafayette realize, if something was going to happen, he would have to be the one to do it. He tried to think of ways to change hearts and minds. Maybe he could take the lead in proving to the world how abolition would help the economy.

Believing his American friends would eagerly accept this challenge, on February 5, 1783, Lafayette wrote to George Washington that it was time for America to deal with slavery.

Suggesting they should join together in buying a plantation, freeing the slaves and hiring them to work, thus proving to the world abolition could be accomplished without massive bloodshed, Lafayette pointed out if Washington got involved, he could motivate many other people. After all, Washington was one of the most deeply respected leaders in the world.

Lafayette wrote, *"If you do this, then other people will follow your example. Once we succeed doing this in America then we can try elsewhere. This might sound crazy but I'd rather be crazy this way than considered wise the other way."[181]*

When Washington received the letter, he took some time to think about Lafayette's idea. It did sound crazy but stranger things had happened. Maybe they could work on it together.

Washington wrote back to Lafayette, *"Your idea is striking evidence of the goodness of your heart. I'd be happy to join you in doing it. Let's discuss it the next time that you visit."[182]*

Washington invited him to visit Washington's home at Mount Vernon so they could discuss plans. In those days, since traveling from France to America took a long time, it would be an entire year before Lafayette reached Mount Vernon.

One year later, Lafayette received a very warm welcome in America. He was invited to speak to Congress, which was rather small at that time since America was still only a few states. Congress honored Lafayette's service by appointing him the official ambassador of America to the world. They also presented him with a sword engraved with the portrait of a young Lafayette in military uniform.

Lafayette appreciated their efforts but there was something else that meant a lot more to him. During the darkest days of the war, he had watched the American people suffer for their freedom. Really nice homes had been seized as headquarters for British officers. Many women had been forced to watch their homes be burned down by the British in retaliation for their men serving in the military.

Now that the war was over, Lafayette saw America rebuilding from the ashes. He wrote to his wife about watching America rise again. *"Every step I take in this country brings me new joys. The nation is happy, peaceful, prosperous. The houses I saw burnt—are now rebuilt. Abandoned properties are now occupied. Everything seems to be on the way to complete recovery. I can only hope that my presence here will help them."*[183]

On that trip, he saw something else in America that deeply troubled him. Lafayette had believed that after America won the Revolutionary War, slavery would be destroyed. Yet as he visited his friends: George Washington, Thomas Jefferson and James Monroe, he was deeply grieved to realize the men, who had brought freedom to the nation, would not bring it to their own plantations.

Lafayette wrote to his abolitionist friend John Adams, how grieved he was that the land of the free and home of the brave had become a nation controlled by slaveowners. That this horrific *"crime"* was being *"Perpetrated under our dear flag of liberty."*[184]

Something had to be done. Lafayette returned home to France, determined to take the lead in destroying slavery around the world.

Knowing his name was revered around the world, Lafayette wrote to his friend Alexander Hamilton, who was working on the newly formed New York Antislavery Association. Lafayette asked for his name be added to the membership of that organization in hopes he could motivate other leaders to join in the fight. This organization would succeed in abolishing slavery in New York.

Meanwhile, Lafayette put his other idea into motion. He purchased a plantation in the French colony of Cayenne. (Cayenne is known today as French Guiana.)

Shortly thereafter, he wrote to Washington that he had set the plan in motion and was going to free the people. Making it clear his goal was bringing *"Our Negro brethren"* into full status as citizens, Lafayette wrote, *"God grant that it may be propagated."*[185]

Washington wrote back saying he hoped others would be inspired by that courage. *"Would to God a spirit like that would take hold of the people of this country. But I despair of seeing it. There have been a few abolition petitions presented to the Virginia Assembly but they were barely noticed."*[186]

Lafayette's dream came true. The story would be told and retold and passed down through future generations about the day freedom came to Lafayette's plantation. First, all the things which had been used to punish the slaves, were gathered into a large pile and turned into a bonfire. Then wages were paid to the workers, according to the amount of work they did.

Many years later, Prime Minister Lord John Russell (1792-1878) would describe how Lafayette had made a difference.

"General Lafayette—the consistent friend of human freedom—made a practical experiment of emancipation as early as 1785. In the French Colony of Cayenne, most of the soil belonged to the Crown and he was able to obtain it on easy terms. He expended $30,000 in purchasing land and slaves. He employed a friendly and wise gentleman to take management."

"The first thing the agent did when he arrived in Cayenne was to call the slaves together, and in their presence burn all the whips and other instruments of punishment. He informed them that General Lafayette had bought them for the purpose of enabling them to obtain freedom. He then stated to them the laws and regulations by which the estate would be governed and the financial advantages that would be granted according to degrees of industry. This stimulus worked like a charm. The energy of the laborers doubled."

Yet just as Lafayette's dream began to unfold, it would be disrupted by the French Revolution.

France was on the verge of a major crisis. The nation was running out of money. Lavish royal spending combined with shrinking revenue from an outdated system of taxation had led to an empty national treasury. Running out of options, King Louis called his advisors together to figure out a solution.

The main problem was the outdated, inefficient tax system. At the time, the nobles were not paying any taxes. Taxes were paid by the common people, who struggled under the system's heavy burden. What taxes were collected, weren't making it to the national treasury but were getting devoured along the way.

The King's main advisor, Calonee, suggested a complete overhaul of the taxation system. New taxes would be levied on the nobility. The poor would be freed from the oppressive laws such as the *Corvee*, which required them to work without pay on repairing roads. Calonne proposed that new tax dollars be used to create more jobs and the *Ferme* be abolished to stimulate the economy.

These proposals made sense to King Louis. Proceeding with the first step, he summoned the *Assembly des Notables* to review, consider and hopefully approve the new proposals. Invitations were sent out to the 144 nobles who comprised the Assembly. One of those nobles was Lafayette.

Lafayette answered the summons and traveled to Paris to attend this historic meeting. The meeting lasted for weeks as the noblemen debated the new national proposals.

Everyone knew something had to be done but what exactly was the best solution? While they were willing to remove trade barriers like the *Ferme*, they didn't like the idea of having to pay taxes. Nobles had never paid taxes in France before, why start now? All the proposals were overwhelmingly rejected.

That frustrated the French people. Everyone knew change was needed. The system was starting to crumble. Something was going to have to replace it.

Lafayette took the lead in calling for reforms, daring to stand up in the *Assembly* to complain of how the royal family was depleting the national treasury by spending lavishly while the country went broke. He demanded an investigation into allegations that the royal family was illegally profiting from shady real estate deals.

"Why are they buying more property for the King when most people agree he should be selling the surplus lands he already owns? Millions in tax revenue is being wasted when it is the fruit of sweat tears and blood of the people who are sacrificed to amass the sums so carelessly wasted."[187]

Lafayette reminded everyone that the real problem was the *Ferme's* crushing grip over the national economy. It was time to destroy the trade monopoly and allow the free flow of goods to raise tax revenue.

In a powerful speech, Lafayette described how the common people were being crushed. It was only a matter of time until they revolted. Something had to be done. *"The time has come for us to beseech His Majesty to convoke a National Assembly."*[188]

Gasps were heard across the room. Had he really just said that? The closest thing to a National Assembly in France was the Estates General which had not convened in over a hundred years. It was to France what Parliament was to England and Congress to America: a legislative assembly representing the people. It was comprised of three groups: the nobles, the clergy and the common people.

To call for a session of the Estates General was extremely controversial. Lafayette had gone way beyond that in calling for a National

Assembly that fully represented all of the French people, not just those at the top of society.

Lafayette knew the Estates General kept power in the hands of a few people. Legislation was passed by a two-thirds vote. Thus, any two groups, such as the nobles and the clergy, could unite together against the third group: the common people. One group could force through any legislation they wanted without the common people having veto power. That's why Lafayette had called for a true National Assembly where the common people would have an equal vote. Facing a national crisis, the King decided instead to summon the Estates General.

Thomas Jefferson, who was still living in Paris at the time, showed up to watch the pageantry unfold. The Estates General convened. The King entered in a magnificent procession and took his seat. The nobles, clergy and representatives of the common people took their seats.

Lafayette took his seat among the nobles, then used his position to call for more reforms. He proposed abolishing slavery, granting religious freedom and revising the harsh penal code. Once again, his proposals would be blocked, but this time the voice of the people would be heard.

Outside the legislative chambers, the people on the street were getting uneasy. Food shortages were still spreading. Mob violence was growing. Crime was rising as desperate people saw no other option than breaking into the wealthier homes, looking for anything they could steal and sell for food.

Even Thomas Jefferson's own home in Paris was vandalized several times. Jefferson wrote to his friend John Jay in America, *"Paris is in constant danger of riots over lack of bread. Patience is worn threadbare. Everyone is expecting civil war."*[189]

The storm was coming. While everyone knew things were growing worse, no one had any idea how bad they were going to become.

Realizing something had to be done quickly, Lafayette decided it was time to draft his own version of the *Declaration of Independence*.

Sitting down with his friend, Thomas Jefferson, he wrote what would become one of the most important documents in French history: *Declaration of the Rights of Man.*

Meanwhile, trouble was brewing in the Estates General. The Third Estate, which represented the common people, began to protest the possibility that they could be outvoted by the clergy and nobles.

How could this assembly truly represent the interests of the nation, if the majority of the nation could be outvoted so easily? Motions were made to create a real National Assembly. The Third Estate voted to hold a true National Assembly meeting, inviting the clergy and nobility, but insisting that this time, all voting would be held on equal terms. That night they adjourned, planning to meet again in the morning.

The next morning the Third Estate showed up to find the doors locked. Soldiers stood in front of the entrance, blocking all access. No one was allowed to enter the building where the Estates General met.

Word on the street was that the King was displeased at the potential shift in the balance of power. Orders arrived from the King telling them to disperse and return to their place in the Estates General.

They disobeyed. The Third Estate found a nearby tennis court and held their National Assembly as planned even though they had to stand in the rain as they worked. They took an oath to continue meeting and representing the needs of the nation, regardless of the King's orders. Then they sent word to notify the King that the National Assembly had been formed.

The King responded by visiting their tennis court and ordering them to disperse. Several of the men quietly obeyed and walked out.

Then a gentleman named Monsieur Mirabeau stood up and shouted, *"We are here by the authority of the people; only the authority of bayonets can remove us."*[190]

The rest of the assembly cheered as hundreds of years of feudalism in France was overturned in one day. The King looked around the room and decided it was time to leave.

The next day Lafayette showed up at the National Assembly, took his place in defiance of the King's orders and presented his *Declaration of Rights.*

Destroying the ancient system of feudalism, it declared that everyone was born free and equal, slavery was abolished, religious freedom protected and government spending limited.

Most importantly, to limit the abuse of power, all government officials were to be held directly responsible by the people. Taxes could only be raised by consent of the national legislature. Women were given the right to vote.

This proposal had come just in time. Riots were breaking out all over France. The city of Paris had descended into chaos as desperate people resorted to desperate measures. As the National Assembly discussed Lafayette's proposal, they were interrupted by news that a violent mob had stormed the Bastille.

This was a major turning point in French history. For hundreds of years the Bastille had been a symbol of oppression. That was the prison where the King punished anyone that he didn't like. There was no trial or appeal. Now a furious mob was ripping it apart. The warden of the Bastille was murdered in cold blood. As the city of Paris descended into violence, the King turned to desperate measures.

The French military was ordered to march on Paris and restore order. Thousands of troops quickly converged on the city in a massive show of force. Meanwhile, the King went to the National Assembly and said what no one ever thought they would hear him say. He recognized their authority. Promising to honor their decisions, the King asked for their help in restoring order to Paris.

There was only one person who could. Lafayette was the only official the nation trusted.

The National Assembly turned to Lafayette, begging him to take command of the French military and offering an extravagant salary of £220,000 plus a generous spending allowance.

Lafayette accepted the job but rejected the salary, saying France needed to cut its spending. Then leaving his place in the legislature and going back to his position as military general, he made it clear, *"All I want for the nation is liberty, order and a good constitution."*[191]

Once again his military skills would save many lives. Lafayette quickly organized the French military into providing law enforcement that kept the peace as much as possible. Still he could see that he was quickly losing control. With the people starving, it was only a matter of time until France descended into complete anarchy.

Trying to save as many lives as possible, he sent the military to open up granaries and distribute the food to the people. That calmed matters down temporarily but there was a bigger problem.

After suffering for hundreds of years, anger on the streets had exploded. People were taking matters into their own hands and confronting corrupt government officials. Mobs were hunting down and attacking anyone believed to have taken advantage of the people.

Lafayette desperately tried to maintain law and order. He was able to rescue several government officials just as angry mobs tried to kill them. Yet he could see the violence was out of control.

He described: *"The people are insane, drunk with power. They will not listen to me forever. Eighty thousand people have surrounded the Hotel de Ville and cry out that we are lying to them."*

"The minute I am gone, they lose their minds. I reign in Paris but I reign over an angry population aroused by evil conspiracies."[192]

Lafayette continued working day and night, doing whatever he could to protect the nation. He returned to the National Assembly in time to cast his vote in favor of destroying feudalism.

Many reforms were made. Serfs were freed. Taxes reduced on the poor. Nobles were taxed for the first time in France's history.

Lafayette also proposed the National Assembly divide itself into two chambers like the U.S. Congress.

Having just read a copy of the Constitution of the United States that had mailed to him by George Washington himself, Lafayette wrote to Washington, *"The spirit of liberty is gaining in this country."*[193]

The Assembly listened to Lafayette. They approved his *Declaration of Rights* over the objections of the King, but they also watered it down. They removed women's rights, complaining those reforms were too radical.

Still Lafayette kept pushing the National Assembly to draft a written Constitution to govern the nation. He could see the nation wanted change, but if they were going to transition from feudalism to democracy then they needed to define the laws.

The National Assembly listened to him but procrastinated. They took their time until time ran out. By the time that they finished their new Constitution and Lafayette celebrated the realization of his life-long dream for democracy to come to his nation, that dream had turned into a brutal nightmare.

While Lafayette was busy on the streets trying to keep the peace, the wrong people took over the National Assembly. They changed the laws in France to seize control of the nation.

Abolishing the Constitution and declaring themselves as the supreme authority in France, they turned democracy into a dictatorship.

To everyone's surprise, the one person who would stand up to them would be the King himself. Using his veto power granted by the Constitution, King Louis tried to restore the balance of power by vetoing the power grab made by the National Assembly.

The Assembly ignored the King and pressed forward towards a dictatorship. Lafayette heard what was going on and quickly returned to the National Assembly to confront the troublemakers.

In one of the most powerful speeches of his life, Lafayette stood up in the National Assembly and told them what they didn't want to hear. He named names of troublemakers who were trying to take advantage of the revolution.

Lafayette demanded that the National Assembly stop them before they did any more damage. Didn't they understand that the whole point of writing a constitution was to follow it?

Boldly he stared them down and said, *"I blame you gentlemen. You cannot deny that the Jacobin group is causing all the disorder. Blinded by ambition they are usurping the power of the French people. I accuse them openly."*

"Rejoicing in disorder, they would overturn our laws and the authority that the people have established. I implore the National Assembly to arrest and punish the leaders of violence for high treason against the nation."[194]

At this point people were wondering why Lafayette didn't just forcibly eject all of the troublemakers. Believing in the principles of democracy, Lafayette refused to seize absolute control over the nation. He respected the balance of power and authority of the National Assembly even when that authority would be used against him. Preferring the messy transition to democracy to any type of dictatorship, Lafayette refused to seize control.

Watching Lafayette's speech was the new U.S. Ambassador to France, Gouverneur Morris. He was the Founding Father who had dared to stand up for abolition at the Constitutional Convention. (By this time, Thomas Jefferson had returned to the United States to serve as the first ever Secretary of State under President Washington.)

Morris watched Lafayette give one of the most powerful speeches he had ever seen. He later described in a letter home that Lafayette's speech had shocked the Assembly.

Mouths dropped open as Lafayette boldly told them what had to be done to save the nation. But they would fail the nation as Lafayette returned to the streets to keep the peace. Morris accurately predicted that Lafayette's speech would turn the Assembly against him.

Meanwhile, Lafayette was still trying to protect the people of France. While he had command of the French military and could move troops around the nation as he saw fit, he could also see the bigger picture.

The military was spread thin, trying to keep the peace around the nation. This left the borders of France unprotected. The possibility of foreign invasion was growing as the world knew about the chaos and violence in France.

Lafayette worried that foreign armies might sweep through the borders of France and easily subdue the nation. So he traveled away from Paris to secure the borders, not realizing the real danger the nation faced, would come from within.

With Lafayette many miles away from Paris, those trying to seize control, saw their chance. Spreading rumors in the streets, they stirred up the common people into a violent frenzy. Stores were looted. Churches were ransacked. Museums destroyed. Anything of value was stolen. Anyone who tried to intervene was murdered.

Watching this happen, Ambassador Morris wrote in his diary: *"The murdering continues all day. This afternoon they announce the murder of priests who had been shut up in the Carmes. They then go to the Abbaye and murder the prisoners there. This is horrible that one of the finest countries in the world should be so cruelly torn to pieces."*

Morris wrote to Thomas Jefferson that the unrestrained violence was growing. *"Thousands have perished in this city."*

Things were getting so bad that even if Lafayette himself returned to Paris, *"He would be torn to pieces."*[195]

As Paris descended into chaos, the royal family tried to escape. They were caught only miles from the French border and returned to the palace.

The failed escape attempt turned the people of France against the royal family, as they felt abandoned by their leadership.

Blaming the royal family for all the problems of the nation, one thing led to another. The rising pressure violently exploded.

Angry mobs descended on the French palace. The royal family watched in horror as their bodyguards and personal servants were brutally murdered right in front of them. The King and Queen were dragged from the palace and locked up in prison.

That shocked the world. When news reached Lafayette, he was too far away from Paris to intervene. Still he ordered the military to quickly return to Paris and save the lives of the royal family.

Those orders were disobeyed. Lafayette watched in horror as the men, who had fought side by side with him, turned against him. As the army mutinied against his orders, Lafayette realized there was no way he could save his nation from the bloody rampage.

The leaders of the French Revolution declared themselves the new government of France, turning the National Assembly into the National Convention.

They put the King and Queen on trial for treason. Both were found guilty and publicly executed. Many other government officials and wealthy aristocrats were also killed.

One of Lafayette's closest friends tried to escape the country, only to be dragged out of his carriage and murdered in front of his wife. She was sent to the guillotine.

Lafayette was blamed for the violence.

The leaders of the revolution worked behind the scenes to make Lafayette the scapegoat of the nation. He was charged with treason, even while he was hundreds of miles away trying to save the nation.

Hundreds of miles away, news reached Lafayette of the bloodshed in Paris and the warrant issued for his arrest. He was left with no other choice but to run for his life.

Running out of time, Lafayette ran for the border, hoping to catch a ship and escape to America.

He didn't make it. Being too famous to hide, he was captured and sent to a prison in Austria.

PERSECUTION COMES

When news of his arrest reached Paris, the leaders of the revolution turned France against Lafayette. Sculptures of Lafayette in Paris were destroyed. All of his property and assets were seized by the French government.

Watching in horror, Ambassador Morris wrote home that Lafayette had become: *"The most hated man in Europe. He has nothing to hope for, having been crushed by the revolution that he put in motion. He lasted longer than I expected."*[196]

Lafayette's wife Adrienne was also arrested. As the bloodshed of the French Revolution continued, Ambassador Morris worried about Lafayette's wife. Being locked up in jail was the first step towards being sent to the guillotine.

Every day he saw angry mobs murdering people in the streets. One day the victims were priests. The next day it was merchants accused of withholding food from the people. There was no mercy. With aristocrats being the primary target, he worried Adrienne might be the next victim.

Ambassador Morris quickly rushed to the prison where she was being held. He told French authorities that if anything happened to Lafayette's wife, America would hurt anyone who touched her.

Then knowing bribery might be the only option for saving her life, Ambassador Morris personally borrowed a line of credit for £100,000 to help her. He also wrote letters to America to raise more financial support for her.

Many others, including Washington, sent money to Adrienne. They tried to get her released from prison but she would remain in prison. Her life would be spared but many of her family members, including her mother, were killed.

George Washington also tried to get Lafayette released from prison, but Lafayette's role as a revolutionary had made him feared by world leaders. They wanted him locked up in prison, lest he bring turmoil to their nations as well. The Emperor of Austria made it very clear, he believed Lafayette was personally responsible for the deaths of the French royal family.

Lafayette was treated as a criminal and locked in solitary confinement. That gave him a lot of time to think. He thought about his wife and children. Were they safe? Had they survived the violence sweeping the nation? Knowing that aristocrats were being dragged from their homes and murdered in the streets, all he could do was pray his family was safe.

He also thought about the plantation he had bought. What had happened to the freed slaves? Lafayette wrote to his wife, *"I don't know what happened to my plantation at Cayenne but make sure that the Negroes who cultivate it shall preserve their liberty."[197]*

When that letter reached his wife, there was nothing she could do. Their property and assets had been seized by the French government and sold to fund the revolution. The slaves that they had freed had been forced back into bondage.

The plantation and the workers on it, had been sold with the proceeds going directly to the troublemakers that Lafayette had tried to stop. Adrienne was powerless to save them. The only thing she could protect was their children. She hid the children with relatives, while she was trapped in prison for the next several months.

(Prime Minister Lord John Russell said the workers on Lafayette's plantation in Cayenne were later freed in 1793 when France abolished slavery.)

The person who would figure out how to rescue Adrienne would be future President of the United States, James Monroe. He was sent to France, as the new U.S. Ambassador, when Gouverneur Morris returned to the United States to serve in Congress.

When Monroe arrived in France, he arrived with a plan. Using creative diplomatic strategies, he succeeded in getting many people released from French prisons, starting with Adrienne.

The way he did was by having his wife Elizabeth make regular visits to the prison to visit Adrienne, bringing her food and supplies. This brought a lot of attention, as people watched Elizabeth Monroe's fancy carriage traveling to and from the prison.

People began talking about how unjust it seemed for Lafayette's wife to be imprisoned with no official charges. The more people talked about it, the more it embarrassed the government. The tide of public opinion turned, enabling Adrienne to be released and allowed to flee the country.

Lafayette and his family had already been granted U.S. citizenship in honor of his military service. So Monroe provided Adrienne and the children with U.S. passports so they could freely travel.

Adrienne sent their son to America to receive an education under the protection of their friend George Washington. She did not send their teenage daughters because she had something else in mind.

By this time, Lafayette had been suffering alone in a prison cell for over three years. No one had helped him. Even with all of the powerful friends that he had, no one dared to call for his release.

Lafayette was seen as a danger to the world. He was blamed for all the bloodshed of the French Revolution. Many people believed that if he had not pushed so hard for reforms, none of the horrible things would have happened. Most of his friends had turned against him.

No one dared to stand up for Lafayette, except the one person who loved him the most. When the world turned against Lafayette, the one person who stood by his side, his wife, would also be the person who secured his release.

Adrienne took their two teenage daughters and traveled to Austria where Lafayette was in prison. There she petitioned the Emperor of Austria for the honor of sharing her husband's prison cell.

At first the request was denied because prison was no place for a woman raised in luxury.

Adrienne refused to accept no for an answer, saying she missed her husband and wanted to spend time with him. Her wish was granted but the supplies she had brought would be confiscated.

While this was going on, Lafayette had no idea what had happened to his family. One afternoon, he was lying on the floor of his cell, trying to survive another lonely day in solitary confinement, when suddenly the door to his cell opened. There stood his wife, smiling at him.

Lafayette blinked his eyes, thinking this must be a hallucination. There's no way his wife could actually be *standing right in front of him!* But she was. Tears came into his eyes as he held her for the first time in years.

Adrienne was allowed to move into Lafayette's cell while their two teenage daughters were given a different cell next door. They were allowed to take their meals together as a family, which was a luxury they had not had for years.

Lafayette's spirit would be revived. While his health had suffered from the harsh prison conditions, now he had all the time in the world to enjoy his family. After years of serving far away, this was a dream come true. It was also a brilliant strategy by Adrienne.

She created a public relations nightmare for the French and Austrian governments. News quickly spread around the world that Lafayette's wife and children loved him enough to share his sufferings.

Adrienne's health suffered but she started writing letters to all her powerful friends asking for help. Her voice was heard. Everyone began talking about how delicate women would not be able to endure harsh prison conditions for long.

Pressure rose for their release. Lafayette's situation was discussed and debated on the floor of Congress, until it was decided to send the former Ambassador, now Senator Gouverneur Morris, to Austria to negotiate on their behalf.

Morris went to Austria only to realize that the King of Austria was not in a hurry to do anything. Morris had to wait a full three months before getting a diplomatic meeting with the government.

Then the Austrian government refused to recognize America was an independent nation. Morris ran into one obstacle after another, finally realizing the Emperor of Austria had no intention of releasing Lafayette.

While this was going on, Napoleon was rising to power in France. Napoleon had begun his military career by serving in the French military under Lafayette's command. For many years Napoleon had admired Lafayette, as Napoleon moved up in the military chain of command. When Lafayette was stripped of his authority and forced to flee for his life, Napoleon quickly rose to command the military.

Napoleon proved himself to be a powerful military leader, winning many battles. He crushed the mob violence of the French revolution, restoring law and order on the streets. He also found ways to finance the French government by invading the rest of Europe, seizing new territory and gaining the spoils of war. Napoleon became a hero to the people of France as he restored their national pride and dignity.

And to the shock of the entire world, Napoleon would be the one to free Lafayette from prison. While Lafayette was sitting in an Austrian prison, Napoleon invaded Austria. The Emperor of Austria was forced to sign a peace treaty, surrendering territory and releasing Lafayette from prison.

It was a bitter irony. After all that Lafayette had done to bring democracy to the world, nothing short of a dictator would grant his freedom. The Lafayette family was released from prison under the protection of Austrian soldiers. They went to the American Embassy in Germany to start their new life.

LAFAYETTE'S LATER YEARS

Adrienne had family in Germany that took them in for the next several months, while they figured out what to do. It had been a long five years in prison for Lafayette and two years for his wife.

Poor sanitary conditions in the prison had caused both of them to become very ill. Adrienne was suffering from a severe skin infection. While she received the best of care in Germany, her health continued to deteriorate. Lafayette did everything he could for her, only to feel helpless, watching her suffer.

How could he take care of her with all of their resources taken from them? Knowing he had very few options left, Lafayette wrote to his friend George Washington, asking for help in getting to America. Maybe they could start over in the land of the free and home of the brave. He would receive a reply that he was not expecting.

Washington wrote back that it was not a good time for Lafayette to come to America. The small nation was struggling with threats from both inside and outside.

At that time France still occupied the Louisiana Territory west of the Mississippi River in America. Napoleon had sent troops to occupy New Orleans. Would Napoleon cross the Mississippi and invade the U.S.? Napoleon had already successfully invaded the Netherlands, Switzerland, Germany and Italy and was pressing towards Russia. Just how far was he willing to go to conquer the world?

Washington wrote to Lafayette that America was facing the possibility of war with France. Things were getting so serious, Washington had been dragged out of retirement and ordered back into commanding the American military by the current U.S. President John Adams.

Everyone knew Napoleon was a brilliant military leader capable of using unconventional warfare tactics.

There was a real possibility he might arm all the slaves in New Orleans and utilize them to invade and subdue the nation.

Washington wrote to Lafayette that if he came to America it might upset the delicate balance of diplomatic relations. Would it be the last straw that insulted Napoleon and turned France against America?

What Washington didn't tell Lafayette was that America was trying to buy the Louisiana Territory from Napoleon.

Once again, Lafayette would put America's interests ahead of his own. Following Washington's request, Lafayette stayed in Europe and tried to rebuild his life.

Once again his wife would come to his rescue. While Lafayette had spent his life trying to change the world, Adrienne had been managing their property and assets. She was a talented money manager who knew how to run a business. Her wisdom had kept them very prosperous, until their assets were seized. Once again she would figure out how to do the impossible.

She returned to France and asked for the honor of meeting with Napoleon. Pulling in every favor she could, she negotiated directly with Napoleon himself.

First, she had to get permission for them to return to their home in France, since Lafayette had been stripped of his French citizenship. Cutting a deal with Napoleon, Adrienne promised Napoleon that if they would be allowed to return to their home in France then they would not interfere with his dictatorship.

Napoleon agreed. He needed powerful friends like Lafayette. He gave them their own property back and offered Lafayette a job as the French Ambassador to America.

While this was the opportunity to move to America that Lafayette had been waiting for, he would decline the position. As a matter of principle, Lafayette refused to have any involvement in Napoleon's dictatorship even though he could have been richly rewarded.

Lafayette and Adrienne moved back home to France. Then she went to work on their dire financial situation.

By studying the French laws and negotiating with Napoleon, Adrienne was able to reclaim most of the property that had been taken from them.

In exchange for Lafayette refraining from any political involvement, Napoleon granted them the return of most of their assets. Lafayette's name was also removed from the list of political enemies. Full French citizenship was restored for their family.

Adrienne didn't stop there. She organized tenants to work their properties for a share of the harvests. This helped provide jobs, feed families and restored a steady income to their family.

Within two years of leaving prison, they were making an annual income of over £200,000. Even Napoleon himself was impressed by Adrienne saying she had *"A great deal of character and intelligence."*[198]

Something else returned to them. Their son, who had been sent to America, had gotten an education at Harvard. When it was safe, he returned home to France. Following in his father's footsteps, he served in the military and later served in the French legislature, the Chamber of Deputies.

Lafayette and his wife settled down into a quiet retirement. Just as they were getting comfortable, a special invitation arrived from the U.S. President Thomas Jefferson.

America had just successfully negotiated the Louisiana Purchase. Napoleon had been paid $15,000,000 in exchange for all French claims to territory in the United States. Now someone who understood French culture was needed to govern this new territory. Since most of the people living in the Louisiana Territory were still loyal to France, would Lafayette like to serve as territorial governor? The position came with a generous salary and land grant.

The offer was very tempting. Living in America had been Lafayette's dream for many years. But there was a lot to think about. If he accepted, what would happen to the rest of his family in Europe? Would they be retaliated against? He had turned down Napoleon's invitation to serve in the government, with the excuse of retirement. If he were to come out of retirement to work this job, it would insult Napoleon.

Lafayette also had family serving in the French military. Would they be punished for his decision? Then there was his dream of bringing full democracy to France.

In the back of his mind, as impossible as it seemed, Lafayette felt that maybe someday, there would be another chance for France to choose democracy. When that day came, they would need his leadership. He wanted to stay there to be in the right place at the right time. Believing this, made Lafayette decline the job opportunity of a lifetime. And once again his feeling would be correct. Napoleon's reign would not last forever.

While Napoleon was one of the most successful military leaders in history, he would be destroyed by one fatal mistake. After conquering much of Europe, he decided to invade Russia. He had no idea that the freezing cold weather would crush his forces. Russian forces simply retreated as Napoleon advanced.

The Russians set fire to their own buildings, leaving nowhere for Napoleon's troops to find shelter from the bitter cold. The fire burned everything. Entire forests were left in ashes. Soon French forces were overcome by the cold weather with no trees or natural resources to shelter them. By the time that the French reached Moscow, their army had been drastically weakened. That would give the rest of Europe time to unite against him.

At the council of Vienna, the world's leaders told Napoleon to retreat or face consequences.

Soon after, a world military comprised of British, Spanish and other forces swept into France and crowned Louis XVIII as King. Napoleon was exiled.

While all this was happening, Lafayette was busy feeding the people in his community. When his property managers told him there wasn't enough left in his granaries for all the hungry people coming to him, Lafayette moved his assets around to bring more food to the people.

This earned him the gratitude of the local community. It also earned him a seat in the new national legislature.

While the French authorities were not happy to see Lafayette back in the legislature, they could not stop the will of the people, who still considered Lafayette their hero. Once again Lafayette used his elected seat to advocate for reforms. Then he received a special letter.

In 1824, he received a new invitation from President James Monroe. America was getting ready to celebrate the fiftieth anniversary of the *Declaration of Independence.* Would Lafayette like to come as the guest of the nation he loved?

Finally Lafayette could return to America. But what was supposed to be a short tour of America, turned into a year long visit. Lafayette was overwhelmed with invitations for appearances. Everyone was eager to see one of the few remaining generals of the Revolutionary War.

Entire cities from Boston to Philadelphia to New York planned elaborate celebrations. By the time Lafayette reached America, he found himself as one of the biggest celebrities in American history.

There were lavish parties, gigantic parades and enormous crowds. Everywhere he went Lafayette was greeted as an American hero. Many of the military veterans, who had served with him in the war, came to see him. They shared stories of Lafayette's generosity during the war.

At one of the events, a veteran showed up, wearing his military uniform with half a blanket draped over his shoulder. He saluted Lafayette who returned the salute. Then he shared a story.

During the darkest days of the war, when they were suffering in the brutal winter at Valley Forge, he had been posted alone on guard duty. Late one night, he was shivering in the bitter cold with very little clothing on, as the weather dropped below freezing.

Lafayette had been making his rounds, when he saw this soldier standing with his musket. Lafayette told him, *"Go to my hut. You will find clothing there and a blanket, which after warming yourself, you will bring here. Meanwhile give me your musket and I will keep guard."*

The soldier said, *"I obeyed and after returning to my post refreshed, you cut the blanket in half, kept half and gave me the other. Now here is the half-blanket that you saved my life with."*[199]

Lafayette was deeply touched. Throughout his tour of America many more stories like that would be told of his kindness.

The city of Boston prepared a special surprise for him. Early one morning, he was awakened by the sound of a military group.

When he went to the window and looked outside, a group of volunteers was reenacting the American military unit that he had once commanded. They had even gone to the trouble of recreating the exact uniforms of the unit that Lafayette had once bought when the military lacked warm clothing.

Lafayette loved it, applauding the performance enthusiastically. After their performance, the crowd mobbed him. He shook hands and signed autographs. Someone commented how they were surprised Lafayette was so fluent in English.

Lafayette replied, *"I am an American after all—just back from a long visit to Europe."*[200]

Lafayette was honored by America in many ways, including financially. President James Monroe asked Congress to grant him a large piece of property in Florida and an annual income from government bonds.

Congress threw a lavish celebration to remember the sacrifice he had made. At the party, Speaker of the House Henry Clay raised his glass to toast Lafayette as *"The apostle of liberty whom the persecutions of tyranny could not defeat, whom the love of riches could not influence and whom popular applause could not seduce."*[201]

Lafayette replied by raising his glass for a toast. *"America had always saved us in times of storm; one day it will save the world."*[202]

While Lafayette's visit was only supposed to be a few weeks, it stretched longer and longer. America didn't want him to leave. Town after town demanded the chance to meet him. Months passed by as Lafayette traveled across America.

He stayed through the winter in Washington, D.C. and watched one of the most unusual presidential elections in history.

In 1824, four candidates ran for President: Henry Clay, William Crawford, Andrew Jackson and John Quincy Adams (son of John Adams). While Jackson had won the popular vote and gotten the most electoral college votes, there wasn't enough of a majority in the electoral college to elect the President.

That left the House of Representatives to choose the President under the terms of the Constitution. This was an interesting choice. Adams was the Secretary of State. Clay was the Speaker of the House. Crawford was the Secretary of the Treasury and Jackson was a respected military general.

While they were all highly qualified there was a big difference. Three of the candidates were in favor of slavery. John Quincy Adams was the only abolitionist.

The House of Representatives convened to elect the President. Under the Twelfth Amendment, this voting was done by state. Each state cast one vote for President, which would be determined by a majority vote of all the representatives from that state.

Once again, the voting was hopelessly deadlocked. The tie-breaking vote came down to the New York vote. The New York delegation split right down the middle, with half for Adams and half against. The tie-breaking vote was in the hands of Congressman Stephen Van Rensselaer.

This was a man of prayer, who took the time to seek God's will before making major decisions. While sitting in his seat, he closed his eyes and prayed for guidance. When he opened his eyes, he saw a piece of paper on the floor with Adam's name on it. Taking that as divine providence, he cast the tie-breaking vote for John Quincy Adams to become the sixth President of the United States.

That shocked the nation. Everyone had expected Jackson to win and four years later he would become President. Now there was someone in the White House that no one had thought would be there. Little did they know how much of a difference Adams was about to make.

Lafayette celebrated his friend's election. He had known John Quincy Adams and his father for many years. Both of them had served as U.S. Ambassadors in Europe and were frequent guests in Lafayette's home.

Lafayette had watched John Quincy Adams grow up from being a ten year old, who went to England with his father, while the peace treaty was being negotiated to end the Revolutionary War, to now being the President of the United States.

Lafayette knew he would do a great job. Yet he had no idea Adams was being put in position to turn the tide of America.

Lafayette continued on his goodwill tour. City after city turned out to honor him. He received so many fan letters that he worried how he would ever have time to answer all of them.

The problem was solved while he traveled down the Mississippi River on a steamboat. The steamboat had problems and ended up sinking into the river. All of Lafayette's belongings were lost, including the bag of letters. He was relieved, knowing now he wouldn't be expected to respond to all those letters.

After spending well over a year in the country he loved so much, Lafayette felt it was time to return home. He just needed to say goodbye to the friends who had helped him.

By this time Adams had been sworn into office as President. He invited Lafayette to stay at the White House to rest from the constant crowds demanding his attention. Adams also took Lafayette for one last visit with old friends Thomas Jefferson, James Monroe and James Madison.

While friends like George Washington and Ben Franklin had already passed, there were still many of the original founders who remembered what Lafayette had done for them. Even John Adams was still alive and thrilled to see his son and his old friend.

They reminisced for hours about the miracle of America. How no one thought they could gain their independence. How America had grown so rapidly, the once struggling colonies were now a powerful nation. Lafayette was touched to see his dream come true. Now it was time for him to bring that dream home to France.

Finally when it was all said and done, the time came for Lafayette to board the ship to return home. Everything shut down in the city of Washington D.C. All government offices were closed so everyone could wish farewell to this hero. Thousands of people showed up at the dock to say goodbye.

President Adams thanked Lafayette on behalf of America for love that he had shown. *"Your name will forever be linked to that of Washington's. As your ship leaves, we will be praying for you."*[203]

Lafayette was deeply touched. Unable to speak for several minutes, he wiped tears away from his eyes, knowing this was the last time he would see America. Then he raised his voice to say to Adams, *"God bless you sir and God bless the American people."*[204]

With that, Lafayette boarded the ship as the thunderous noise of cheering filled the air. He returned home just in time to find his nation in another crisis.

King Louis XVIII had died and his brother King Charles X had taken the throne. Once again France was suffering. The Chamber of Deputies (France's new legislature) was locked into a battle with King Charles X. Freedom of speech was being tightly controlled. The King was dissolving the Chamber whenever they refused to obey his orders. Tensions were rising as the French people began to fight back.

Once again, Lafayette was elected to the Chamber.

The King dissolved it for the third time. This time the people rose up. Mobs marched on Paris, violently seizing control. The King had no choice but to flee as history repeated itself.

When the Chamber of Deputies reconvened, Lafayette found the nation asking for his help. Everyone demanded he take control of the French army to restore law and order.

Once again, Lafayette answered the call of duty. He promised the nation that he would devote himself to what was best for them. *"My conduct at age seventy-three will be what it was at thirty-two."*[205]

This was the chance at reform Lafayette had hoped for. It came with the potential for a repeat of the violent bloodshed of the previous revolution. Yet this time everyone knew Lafayette was the only one who could keep the peace. He was given full dictatorship to rule the country as he saw fit.

Lafayette didn't believe in dictatorship. While the country wanted him to seize control, he used his power to steer it towards democracy.

Yet Lafayette realized that his country was not fully ready for total democracy. France needed a powerful King to keep the peace while it transitioned to constitutional monarchy. So Lafayette selected a fifty-seven year old military veteran, who had lost his father to the guillotine, to assume the throne.

This man did have a hereditary claim to the throne as a direct descendant of King Louis XIII. Lafayette crowned him the new King of France. Then he started drafting constitutional reforms.

Lafayette warned the nation that either, *"Liberty shall triumph or we will all perish together."*[206]

Then he wrote the *Programme* to define the most pressing areas of change for the nation. The new king would have to approve this document to take the throne, thus turning France into a constitutional monarchy. The most important element was that everyone, especially the King, had to recognize the authority of the people.

The justice system was completely restructured to allow judges to be elected by the local communities instead of appointed by the King. Finally the common people were given the right to vote, without any restrictions of property ownership. Lafayette hoped this would keep the power in the hands of the people.

The end result was a bloodless revolution. Government authority passed to the new King while the Chamber of Deputies approved the new reforms. The court system was also reformed to protect the right to due process. No longer could government leaders simply imprison anyone at will.

This was a major turning point in history. The nation celebrated. As Lafayette described, it was the best solution at the time.[207]

The peace did not last for long.

Once again the new King tried to seize absolute power. The Chamber was dissolved. Riots prevailed. Martial law was imposed. Violence escalated. Blood was shed and Lafayette found himself once again trying to save his nation.

This time his strength failed him. At seventy-five years old, his body was growing weaker, leaving him barely enough strength to go about his daily activities while he continued pushing for reforms in the Chamber.

Yet the nation supported him. The people followed him even as his own doctor frantically tried to keep him alive. Slowly his life slipped away. Trying to encourage him, the doctor said he had become a father figure to France. Lafayette replied that it still felt like his words were not being heard.

Lafayette died on April 20, 1834, while holding a picture of his wife. (She had died in 1807 after a long illness). He was buried next to her. Soil from the Battle of Bunker Hill was brought to France and spread over Lafayette's grave, to fulfill his dying wish to be buried at his home underneath American soil. The entire nation of France grieved the loss of their hero.

In the weeks that followed, America also honored his legacy.

Lafayette became the only person in American history to receive the same burial honors as George Washington.

All flags were lowered to half-staff. Congress adjourned to hold a day of mourning. Both houses of Congress were draped in black.

The American military fired a salute every hour from sunrise to sundown, on the day of mourning. President John Quincy Adams gave the eulogy at the memorial service held for the nation's leaders. He celebrated the memory of one of the greatest friends America had ever had. No one else had ever accomplished what Lafayette had.

Lafayette's dream would come true. It would be a long road with many twists and turns but just as Lafayette had predicted, one hundred years after helping America win the Revolutionary War, France would become a full constitutional republic. Dictatorship would be overthrown and full democracy would come.

Yet while liberty was spreading across France, it was shrinking in America. The land of the free and home of the brave had become a nation controlled by slaveholders.

While they were only a small percentage of the population, they were pulling all the strings of the federal government. Lafayette had seen this on his last visit to America. He had been deeply grieved to realize the freedom, he had sacrificed so much for, was slowly disappearing. Democracy was being used to protect slavery.

Lafayette had tried to stop that. He had advocated for abolition as much as he could. In the last years of his life, Lafayette stayed in contact with his abolitionist friend Thomas Clarkson who was working with William Wilberforce.

Lafayette wrote to Clarkson that while he was greatly encouraged by the progress England was making towards abolition, he was deeply grieved at the hardness of heart in his own nation.

"The present system cannot last. The danger is extreme. But you saw the opposition we had to encounter from the French aristocrats of that time when I worked with you on abolition. I was ashamed when procrastination became necessary."

"But I told you that our doctrine of liberty would eventually succeed in destroying the slave trade. As you continue the battle, I pray that God will grant you success."[208]

Just months after being released from the Austrian prison, Lafayette had written letters to Thomas Clarkson, hoping to continue working towards the *"Restoration of rights"* to *"Our Negro brethren."*[209]

And while he was trying to rebuild his life after coming home from prison, Lafayette stayed true to those principles.

At one point when he needed finances the most, some friends in America offered to provide him with a steady income stream by setting up a tile business on land in America that had been given to Lafayette.

He declined that offer. The business plan had been unacceptable because it involved purchasing thirty slaves.

Lafayette wrote to James Madison that he wanted nothing to do with profits earned from the trade *"That I detest."*

Going even further, he told Madison he would only get involved in projects that were *"Productive of freedom."*[210]

His dream came true. Lafayette lived long enough to see the triumph of Clarkson and Wilberforce, when England abolished slavery in 1833. Yet Lafayette still grieved that freedom had not yet come to America.

In the last moments of his life, Lafayette wrote letters to Clarkson on how thrilled he was that England had destroyed the evil trade. He wrote, "*1500 petitions presented to Parliament with 1,500,000 signatures is a glorious event. I'm glad you have lived to witness this event.*"

Lafayette shared with Clarkson his concern over how America was losing the battle on abolition. Disturbed over the direction the leaders of America were taking, Lafayette told Clarkson, *"I would have never drawn my sword for America if I had thought that I was establishing a slave-holding nation."*[211]

Lafayette was not the only one worried about the future of America. As a dark cloud of evil wrapped itself around the nation, the President of the United States was down on his knees, crying out to God to save America.

Day after day, he fasted and prayed that somehow, someway God would use him to destroy that evil.

God heard those prayers.

In the years that followed, the President of the United States and the last living link to the Founding Fathers—John Quincy Adams—would strike the match that would burn into the inferno that destroyed the monster.

Lafayette Images

Bunker Hill attack

Battle of Bunker Hill

Lafayette meets Washington

Lafayette at Valley Forge

Lafayette with Washington

Washington and Lafayette entering New York after victory

Lafayette visiting Washington at Mount Vernon in 1784

Lafayette at home in France with his wife

Lafayette in France during troubled times

Lafayette shaking hands with Uncle Sam shows how America
kept the promise to help France by entering World War I

10

President John Quincy Adams

"Being confident of this very thing,
that he which has begun a good work in you
will perform it until the day of Jesus Christ."
Philippians 1:6 (KJV)

President John Quincy Adams
1767-1848
From Massachusetts

John Quincy Adams hated his job. As the sixth President of the United States, his hands were tied. Every thing he tried to do to help his country was blocked.

When his Presidential term ended, he left office feeling like a failure. At sixty-two years old, it was time to retire. Little did he know that the plan of God for his life was far from over. The most important work he would do was yet to come. From age sixty-two to age eighty he would turn the tide of a nation.

President John Quincy Adams

Adams had grown up living the history that we all studied in school. He was born in 1767 as America prepared for revolution.

His parents, John & Abigail Adams, were well known and deeply respected. He grew up in a home frequently visited by people like George Washington, Thomas Jefferson and Benjamin Franklin.

The Revolutionary War began when the British sent hundreds of men to capture his father's cousin Samuel Adams and friend John Hancock. Of course the British didn't count on Paul Revere waking up the people in time.

Seven year old Adams watched the Battle of Bunker Hill happen. His mother, Abigail, raised him while his father traveled many miles to serve in the Continental Congress. His father worked long hours, handling the paperwork of democracy, trying to get proper funding to the military.

His father also served on the committee that wrote the *Declaration of Independence.* His father even took the time to handwrite a full length personal copy of the *Declaration* to mail home for his mother to see what they had been working on. As a child, Adams got to read that very important document before the rest of the nation did.

Meanwhile, his father had to spend two full days convincing the Continental Congress to approve the *Declaration* and present it to the world.

In 1778, ten year old Adams traveled with his father to Europe, as America began to establish diplomatic relations with other nations. He watched history happen as his father worked with Benjamin Franklin and John Jay to negotiate the peace treaty with England to end the Revolutionary War.

While living in England, Adams visited Parliament, where he sat in the audience and listened to the debates. He saw William Wilberforce powerfully thunder from the floor of Parliament.

Adams learned how to debate from watching both Wilberforce and Prime Minister William Pitt completely shred the arguments of political opponents during intense policy debates.

Adams also traveled to France and stayed as a guest in Lafayette's home. Year after year, young Adams learned political science through real world experience, not textbooks.

As America became a nation, fourteen year old Adams was asked to serve his country on a three year diplomatic journey to Russia. There he worked as a secretary to the U.S. Ambassador. There he also saw how precious freedom really was.

He wrote in his journal, *"In Russia, all the farmers are in the most abject slavery."*

"The country man is attached to the land in which he is born. If the land is sold, he is sold with it. And the nobility have the same absolute power over the people that the king has. This system hurts the king, the nobles, and the people. Already it has brought four revolutions to this country."[212]

Yet Adams also noticed how much the Russians treasured freedom. Some of the peasants would go to great lengths to make enough money to buy their freedom.

Adams wrote, *"I saw a man who paid his landlord 450,000 rubles for the freedom of himself and his descendants. This proves the respect they have for liberty even when they've never known it."[213]*

During the three years he lived in Russia, Adams worked as a translator, copying long diplomatic documents while the young American republic gained the support of Russia.

Adams was fluent in several languages including Russian, French, Dutch and German.

When his international diplomatic service ended, he finally had a chance to attend college. He graduated Harvard with a Bachelor's in 1787 and Master's in 1790. Passing the bar exam in 1791, he began practicing law in Boston.

Meanwhile, George Washington was elected as the first President of the United States. Serving alongside him was the first ever Vice President—John Adams, the father of John Quincy Adams.

In 1794, President George Washington asked young John Quincy Adams to serve his country as the U.S. Ambassador to the Netherlands. Answering the call of duty, Adams returned to Europe for a few years. When that diplomatic service ended, Adams moved back to his home state of Massachusetts and served in the state legislature.

Then Adams was elected to serve in Congress from 1803-1808 as the Senator from Massachusetts.

Throughout that term, Adams did everything he could to serve his country. Thinking outside the box, he never hesitated to stand up for what he felt was right. And that got him voted out of the Senate.

He supported President Thomas Jefferson's policy of purchasing the Louisiana Territory, even though it was hugely unpopular at the time. Since Senators were not elected by the people, but appointed by state legislatures, the Massachusetts State Legislature promptly recalled him home. Adams became a law professor until once again his country asked for his help.

Then Adams served his country as:

- U.S. Ambassador to Russia 1809-1814
- U.S. Ambassador to England 1815-1817
- Negotiated Treaty of Ghent with England to end War of 1812
- Secretary of State 1817-1825 (under President James Monroe)
- Negotiated purchase of Florida territory from Spain in 1819
- President of the United States 1825-1829

As President, Adams focused on improving America through building new roads. He even paid off much of the national debt. Yet the political opposition was overwhelming. Political games railroaded his new improvement projects.

In 1828, Adams ran for a second term and lost the election to Andrew Jackson. Adams left office frustrated at his lack of achievements.

Secretly he confided in his diary, *"No one knows the mental agony that I suffered as President, until I was dismissed from that position by the failure of my re-election."*[214]

With his political service over, he wanted to retire to a quiet life. Adams was happy to leave the public eye. Maybe now he could have more time to enjoy his children and grandchildren. Little did he know that his work was just beginning.

One day he opened the newspaper and recognized his name. His hometown of Boston wanted him to return to Congress. Since the House of Representatives were the only seats in Congress elected by the people, they had put his name on the ballot for the upcoming election.

The nation needed him now more than ever. The people were going to run his political campaign with or without him. Adams was surprised. He decided to wait and see what happened.

In the election, Adams won by a massive landslide. 75% of voters in his district wanted him back in Congress so much that they had elected him without him doing any campaigning.

That made Adams really happy. Knowing he had the people behind him, he wrote in his journal, *"No other previous position, not even being elected President, has given me as much pleasure as being called upon by my district."*[215]

As Adams walked back into politics, he had no idea that what he was going to do would be the most important thing he had ever done for his country. Taking his seat in Congress, Adams hoped to *"Fulfill my destiny."*[216]

There was plenty to keep him busy. James Smithson had donated half a million dollars to the United States for the advance of science and learning. How should the money be spent? Adams was asked to lead a committee to find the right project for the money.

This was a difficult project. Everyone wanted the money. Every state demanded a piece, proposing all kinds of ideas. But Adams had a better idea. He suggested creating the Smithsonian Institute as a place of learning for future generations.

This was a great idea, but Adams still had to fight for it. Several other greedy Congressmen tried to take the money for themselves.

Eventually Adams won and the world's largest museum was born. The Smithsonian Institute still exists today because of Adams.

As Adams served in Congress, there was a lot of work to keep him busy. Yet there was one thing really important to him. Freedom. The most precious part of America.

Growing up during the Revolutionary War, Adams had inherited from his family the love of liberty and hatred for slavery.

His parents had set a powerful example by refusing to own slaves in a time when people thought it was normal.

Their own neighbors, thought they were crazy for paying wages to hire help when labor could be had for free. Yet Adam's parents cared more about truth than comfort.

During the war, his mother had written to his father, *"I wish there was not a slave in the country. It seems so sinful to fight ourselves for what we are robbing from those who have as good a right to freedom as we have."*[217]

Soon as the war ended, his father had written the Massachusetts Constitution in a way that enabled the local state courts to abolish slavery in Massachusetts.

No sooner had independence come to America than the greed of powerful men had seized control, wrapping a choke hold around the federal government.

As this dark cloud of evil swept over the nation, Adams grieved that *"The venom of slavery infected the Constitution until no fumigation could purify it."*

"Like the spokes of a wheel extend out from the axle, slavery became the very axle around which the federal government revolved. It controlled all domestic and foreign policy."[218]

Adams cherished the most precious possession he had: the hand-written copy of the *Declaration of Independence* his father had given him. He couldn't understand how his old friend Thomas Jefferson could have written something so powerful for freedom and then never freed the people on his own plantation.

Yet Adams also saw that the Founding Fathers had left the tools for freedom to the next generation. Someone just needed enough courage to use them.

Adams wrote, *"Jefferson's Declaration of Independence laid the first foundations of civil society; but he never realized that it also laid open an abyss into which the slave-holding planters of this country must fall sooner or later."*[219]

Throughout his diplomatic career, Adams had spent long hours working with the most powerful proslavery politicians.

They had had many discussions on politics and law, yet Adams never allowed them to pollute his mind. He told them exactly how he felt. And as one of the greatest constitutional law scholars ever, he had explained to them just how unconstitutional slavery was.

They never listened. Adams realized that nothing could convince them to free their slaves. Confiding his frustration to his diary he wrote, *"They admit slavery is evil. They blame England for causing it. But at the bottom of their souls is pride in being the master."*[220]

Adams grieved in his diary that the Founding Fathers had no excuse for failing to abolish slavery. He wrote how much he hated that *"The evil of slavery is that it perverts logic and defiles morals with false principles."*

"What can be more false and heartless than this doctrine which makes the sacred human rights to depend on skin color?"

"The compromise between freedom and slavery in the Constitution is cruel and inconsistent with the only principles which justified our Revolutionary War. The consequence has been that slave-holders have ruled America."[221]

Day after day as Adams worked with proslavery politicians, he could see that they would do anything to keep control. There was an evil plan in motion. Civil war was coming. They would push for more and more control until one day the country would lose its independence to keep its slavery.

That's how far they were willing to go. They actually told him that when civil war came, they would appeal to England for protection even though that meant returning America back to the status of a colony ruled by Europe.

Adams could hardly believe what he heard. To deny freedom to others they were willing to lose their own!

Seeing the future was very disturbing. Nothing short of a bloody civil war could save America. Adams shuddered in his diary that one day violence would rip apart the nation with *"Calamity and desolation."*

Yet while he dreaded the horrors of war he knew *"It's end result would be the complete destruction of slavery."*

"If slavery becomes the sword destined to cut the ties of the Union, the same sword will shred the bonds of slavery itself."[222]

Their crushing death grip of control over the federal government was slowly squeezing the life out of the nation.

Someone had to crush the evil power before it destroyed America. But who? And how? And when?

The more Adams thought about it, the more he realized that the power of America was in the hearts of the people. Hadn't a raggedy group of farmers and woodsmen defeated the most powerful nation on earth in the Revolutionary War?

Now they had to resist another tyranny attacking their country. So how could he stir up the country to resist? First he had to get them to see what was happening. He had to expose how the proslavery power trampled across freedom of speech and due process of law.

Yet how could he start a war in the last years of his life? Could his health handle the tidal wave of opposition he would ignite?

Feeling *"Disqualified by my age and infirmities,"* he still decided that he had no choice *"But to fight. I will fight the devil with his own fire."*[223]

Praying and seeking God, *"For a pure and honest purpose,"* Adams wrote in his diary, *"I have a duty to the people. A duty to truth and justice and the memory of my father."*

"I will expose the fraud of slave-holding democracy. I pray for attitude, moderation, firmness, self-control, success and put my trust in God."[224]

The best platform he could get for this message was Congress.

Each member of Congress had regularly scheduled days for speeches and proposals for legislation. Time was given to every Congressman to take the floor and discuss the issues facing the nation. The speeches and votes of Congress were printed in newspapers and read by everyone across the country.

This was a time in American history when the primary form of entertainment was reading. Everyone devoured books, pamphlets, and newspapers.

Even the slaves, who were not allowed to learn to read, would find ways to overhear the master reading the newspaper to his wife. Or they would save discarded newspapers and take them to friends who would read them.

Thus, Adams knew his words in Congress would be heard across the nation. This was the best chance he had of getting everyone's attention.

Something had just arrived in the mail. One hundred and fifty-three people from Worcester, Massachusetts had sent Adams a petition, begging Congress to abolish slavery.

There was a time in history when petitioning Congress was the only way women could be involved in politics. They were not allowed to vote. They could not pick up the phone and call their Congressman. Most of them would never have the chance to meet their Congressman in person. So people would write petitions, asking Congress to address certain problems in the nation.

Once they had their friends and neighbors sign the document, they would mail it to their local Congressman. Then they would watch the newspaper's report on Congressional proceedings until they saw their Congressman present their petition to Congress. If they were lucky, Congress might discuss the issue and maybe even introduce legislation.

The first month of each Congressional session was scheduled for each representative in Congress to take turns presenting petitions from his district. Then Congress would debate potential legislation to address the issues.

While these petitions dealt with a wide variety of issues, the vast majority of petitions going to Congress, demanded that Congress abolish slavery. These petitions were coming from both the North and South.

While most of America hated slavery and wanted Congress to end it, most Congressmen were too afraid to do that. They had taken money from the proslavery people and couldn't afford to lose their political power by speaking up.

Many Congressmen ignored the antislavery petitions and angry letters from voters because advancing their political career required bowing to the evil power.

Adams had already lived a political career way beyond what the average Congressman could dream. What did he have to lose? Having already been President, he wasn't thinking about building a career. What could he do for his country?

Adams looked at this antislavery petition on his desk. Should he present it to Congress? Many sitting Congressmen were slaveowners. They would be deeply offended if anyone suggested it was wrong.

Adams knew that presenting this petition would trigger a violent storm of opposition in Congress. The backlash would be powerful. Could he withstand the pressure? With the deep pockets of slaveowners controlling much of the federal government, anything that stood in their way would be vigorously attacked.

The more Adams thought about it, the more he realized that if he stood up for justice, the opposition could help him. If they fought him, he could show the nation what was happening behind closed doors. How the wrong people were in power. That the system of slavery was going to destroy freedom of speech and every civil liberty in the nation unless it was destroyed first. Maybe if he stood up to this evil, other people would stand with him.

ADAMS TAKES HIS STAND IN CONGRESS

On February 1, 1836, Adams stood up and read the antislavery petition in Congress.

That petition sparked a firestorm of opposition.

Immediately, Congressman Jarvis of Maine jumped up to oppose it. He demanded that Congress refuse to accept any antislavery petitions, since Congress had no right to interfere with slavery.

Mr. Jarvis demanded that all antislavery petitions be ignored and not printed in the official record. He even read a resolution from Maine's State Legislature that any interference with slavery would disturb the peace of the nation.

When Mr. Jarvis had finished, Congressman Glascock of Georgia stood up to say, *"Any attempt to debate slavery in this House is disturbing the compromises of the Constitution, endangering the Union and, if persisted in, would destroy the peace and prosperity of the country by war."*[225]

The debate continued until the House voted to move on to other issues. The petition was *"laid upon the table."* That was the nice way of telling the country that nothing further would be done.

That night, the proceedings of Congress were published in the newspaper. The next morning, people read how Adams had taken a very unpopular stand.

Most newspapers criticized him for doing it. At the time, the income of many newspapers depended on selling ads to slave traders. They couldn't afford to have anyone interfering with their business. So they slammed Adams left and right.

As the tide of public opinion turned against him, Adams refused to back down. In the days that followed, his reputation was viciously attacked.

As the opposition intensified, Adams wrote in his diary,

"Both political parties are slandering me. Everything I do and say is scorned by almost every newspaper. I can see nothing ahead but censure."[226]

Yet Adams was tougher than them. Did they really think they could intimidate someone who had already seen liberty triumph over tyranny in the Revolutionary War? Adams laughed at their scare tactics and forged ahead.

The American people began mobilizing behind him. Letters poured into his office, thanking him for speaking about the forbidden topic. More people sent antislavery petitions.

Adams smiled when he saw them. Now he had more stuff to talk about in Congress. He went back to work, armed with more petitions.

When his turn to speak in Congress came again, Adams stood up and told everyone what they didn't want to hear.

Looking around the room at dozens of slaveholding Congressmen, Adams thundered, *"Slavery is a sin before God!"*

He paused as his words sank into the audience. The Congressmen stared back at him in shock, many with anger glistening in their eyes. Had he actually just said that? How dare he lecture them!

Adams never flinched. His eyes narrowed with righteous anger as he stared right back at them and announced that he was going to war with slavery. *"I will drive it backwards to its corrupt fountain. I will pursue it until it is forced to disappear from this land and the world."*[227]

As Adams preached to a hostile audience, the Congressmen grew enraged. They interrupted him with insults and threats. How dare he interfere with their lifestyle!

Knowing that Adam's words would be published in the newspapers and heard across the land, they had to shut him down before his voice was heard.

So a few weeks later, Rep. James Hammond of South Carolina gave a speech in Congress to remind everyone of how impossible it would be to abolish slavery. He accused Adams of *"Attacking and slandering Southern people."*

He demanded that all abolition groups be stopped because

"The Constitution guarantees slavery. Congress has no right to legislate on it. If it ever did there would be a civil war. Without slavery Southern lands would be useless. They cannot till them for any profit by any other labor."

Making his point clear, Hammond closed his speech by saying,

"I warn these ignorant, infatuated abolitionists that if they fall into our hands they will die a felon's death."[228]

Adams smiled as he listened to the speech. The proslavery power had no idea they were following his plan. By forcing them to discuss the forbidden topic, he was exposing their true character to the nation.

Next time it was his turn to talk in Congress, he presented more antislavery petitions, including one from 158 ladies of Massachusetts. Adams proudly said, *"I have never doubted that females are citizens."[229]*

Proslavery Congressmen were furious that they had to listen to this nonsense. They had to show up because they were only paid for days they were physically present in Congress. So when it was Adam's turn to take the floor, all they could do was listen and object to his obnoxious words.

Henry Pinckney of South Carolina had a better idea, suggesting that Congress create a committee to receive and process all antislavery petitions.

The committee was formed. Petitions were referred. And after a lot of work the committee reported to Congress on May 19, 1836.

Representing the committee, Pinckney explained that during this time Congress had received a total of 176 antislavery petitions signed by 34,000 people, of which 15,000 were female.

This was disturbing. Pinckney asked, *"What good could it do? The government had made a solemn covenant with the slaveholding states."*

"Congress is bound hand and foot. Abolitionism has reached its height. It has begun to go down and will soon disappear entirely if we do not fan the flame ourselves and allow our friends in the free states to fight the fanatics in their own way."[230]

The committee's report presented a resolution to the House *"That Congress possesses no constitutional authority to interfere in any way with slavery."*[231]

Hearing this made Adams jump out of his chair to yell, *"Give me five minutes to prove that's utterly false."*[232]

Adams was silenced by the Speaker of the House.

The House voted on the resolution. It passed by 182 to 9.

Adams wasn't done. Following the vote, he stood back up and roared, *"Mr. Pinckney, are you ready for war?"*

Everything stopped. The Congressmen looked up with puzzled faces. What did he mean by war?

Adams continued, *"Someday your policies will bring civil war to this nation. There will be a great battle fought between freedom and slavery."*

"While you refuse to believe that Congress can interfere with slavery, from the moment that war comes to the South, Congress will be forced to intervene. Can't you see that you are willfully launching these wars and then closing your eyes and blindly rushing into them?"[233]

That did not go over well. Pinckney glared back at Adams and proceeded with presenting his report.

The committee had found that their job was to put an end to any attempt to disturb slavery. Comfort was more important than justice.

Pinckney suggested that Congress pass a new rule automatically rejecting any abolition petitions. Debate over slavery would also be forbidden in Congress. This *"gag rule"* was necessary because they didn't want anyone preaching to them.

Again Adams jumped up from his chair and objected.

He was quickly silenced. The Speaker of the House refused to allow him a chance to discuss why this *"gag rule"* was unconstitutional.

Adams demanded that Congress explain how it could lawfully pass this rule.

He was not allowed to take the floor.

Adams refused to surrender. Staring down the Speaker of the House (and future President) James Polk, Adams demanded a reply.

"Mr. Speaker, am I gagged or not?"

The Speaker moved on to other matters, trying to ignore the issue.

Adams loudly declared, *"I am aware that the Speaker of the House is a slaveholder himself."*[234]

The room glared back at him. No one was supposed to mention that. Adams was told to hush by some very angry colleagues. He sat back down in his chair with another broad smile. This was just the warm up.

The *"gag rule"* made front-page headlines. It only lasted for one session of Congress. Then the Speaker of the House ruled that it had expired.

So Adams brought more antislavery petitions, including one from the slaves themselves.

Once again, Congress got upset at him. The *"gag rule"* was passed again. This time they made it permanent.

They thought no more antislavery petitions would be allowed on the floor of Congress.

Little did they know that Adams had them right where he wanted. He was forcing them to show their true colors to the nation.

Adams knew that the harder they tried to silence him by stomping on the right to free speech, the more the nation would turn against them. He just had to keep pushing their buttons until people realized how far they were willing to go to protect their selfish interests.

When his turn to speak in Congress came again, to infuriate them, Adams defiantly presented more petitions from women.

The rest of Congress was annoyed. Senator Walker of Mississippi gave a speech in Congress because he was *"Pained to see the names of so many sensitive and indelicate American ladies petitioning."*

"Surely they would be much better employed in attending to their domestic duties as mothers, wives, and daughters than interfering in a matter they know nothing about."

"If the ladies would leave us alone, there would be no abolition petitions and the South would not be alarmed."[235]

That angered the ladies of America. Opening their newspapers, they realized the proslavery lobby didn't want women involved in politics. As they kept reading the debates in Congress, they heard more and more infuriating comments such as Rep. Henry Wise of Virginia even suggesting that Congress should provide husbands for these petitioning single ladies to keep them too busy to interfere in politics.

The ladies responded by flooding Congress with petitions.

Thousands of petitions poured into the mail-room of Congress. The Congressional mail clerks worked night and day, trying to process them.

More and more petitions kept coming. As stacks and stacks of petitions piled up in the mail office of Congress, several new full time clerks had to be hired just to process them. Record rooms became filled to capacity until even a large storage room 20' by 30' by 14' packed floor to ceiling could not contain them.[236]

Letters poured into his office thanking him for giving women a voice in politics. Adams was thrilled to see the American people getting involved in the political process. Whenever he got the floor in Congress, he presented their petitions, one by one, to the dismay of many other Congressmen.

Benjamin Howard of Maryland protested that Adams was delaying Congressional business with worthless petitions from females. *"There's more than enough room at home for women to exercise their influence and benevolence. But it's disgraceful to this nation for women to abandon their proper role at home to interfere with politics."*

That inspired Adams to give a long speech defending women's rights. *"How can you say that women are capable of nothing other than bearing children, cooking meals and caring for the home?"*

"Who was it that said that anyway? Isn't Mr. Howard a son, a father and a husband?"

"Then I want him to listen to these petitions from the ladies. While I believe they do have some responsibilities at home, they also have the responsibility of serving their country and it's our job to defend their honor when they do it. Let me give you a few examples."

Next Adams read from a Congressional report on a woman named Deborah Gannett (1760-1827). During the Revolutionary War, when the American military was desperate for soldiers, she had disguised herself and volunteered.

She valiantly fought through very intense ground warfare. As bullets were flying, she was shot twice in the leg. Knowing that she could not go to the military hospital without her identity being discovered, she tried to remove both bullets herself.

One bullet was removed. The other would remain in her leg, preventing her injury from healing properly.

For the rest of her life, she suffered lingering health complications from that injury. After the war, she petitioned Congress to receive the same military pension as other wounded veterans. The pension was denied, until Paul Revere lobbied Congress on her behalf. Then Congress reviewed her claim and changed their mind.

Adams read from the official Congressional investigation, *"This woman defended her country as a common soldier for nearly three years, fighting and bleeding for liberty. Due to the injuries she has suffered and the hardships she endured in defense of her country, Congress grants payment."*

Adams asked, *"Does this report describe the heroism and loyalty of a woman, as a shame to her country? Of course not! I disagree with Mr. Howard on what are women's duties."*

"Are women to have no opinions or actions on the nation's issues? Where does he get that principle from? History?"

"Hasn't he heard of Jael who killed the enemy of her country? Has he forgotten Esther who by her petition saved her country? Our Savior Himself raised Lazarus from the dead at the petition of a woman!"[237]

Then Adams launched into a history lesson from the Roman Empire to the Revolutionary War with example after example of how women had changed the world through political involvement.

Adams roared, *"In America's struggle for Revolution what would the men have been but for the women? When Washington said the troops were destitute of clothing, where did help come from? The ladies of Philadelphia!"*

"Women are not only justified, but display the greatest virtue when they do depart from the domestic circle and enter the concerns of their country, humanity and their God."[238]

Adams words were published in the newspapers causing everyone to discuss the topic of women's rights. Many people were offended at these crazy ideas. Even the newspapers themselves were upset at Adams for interfering with their business.

Since their income depended on selling ads for slaves, they began to fight back against Adams.

The *New York Times* published an editorial slamming Adams as *"The Madman from Massachusetts."*

Trying to discredit his work they wrote, *"Monday is play day for Mr. Adams in the House of Representatives. Time is wasted and business delayed through the waywardness and stubbornness of this strange person determined to have his own way. He ought to retire."*[239]

Adams read that and laughed. Why should he get discouraged when the future of the nation hung in the balance? Ignoring the criticism, he kept coming back to Congress with more strategies up his sleeve.

He knew the South was hiding a dark secret. Certain things weren't supposed to be discussed in polite company. No one was supposed to notice what was standing right in front of them.

Plantation owners often took advantage of female slaves, fathering children with them and then holding those children in bondage. A few freed their children, giving them an education and inheritance. Yet most planters kept their own children and grandchildren as slaves.

Mary Chestnut, wife of the powerful South Carolina Senator James Chestnut, described the plantation lifestyle in her diary.

She wrote, *"God forgive us but our monstrous system is wrong and sin."*

"Like the patriarchs of old, our men live all in one house with their wives and concubines; and the mulattoes one sees in every family partly resemble the white children. Any lady is ready to tell you who is the father of all the mulatto children in everybody's household but her own. Those, she seems to think drop from the clouds."[240]

Many of the Congressmen who fought Adams the most were also guilty of the worst kinds of abuse.

It would later be discovered in a very public scandal that Rep. James Hammond (who led the proslavery movement and ruthlessly fought Adams the hardest in Congress) was himself molesting his four teenage nieces—age thirteen to age nineteen.

Hammond also abused teenage slave girls on his plantation, as young as twelve years old. That was only the tip of the iceberg of the tremendous abuse happening behind closed doors on the plantations.

Adams was deeply troubled that this abuse was being protected by the legal system. If no one else wanted to talk about it, he would make it headline news. He would have to be very careful how he approached this issue, since even mentioning it was considered unthinkable.

On February 6, 1837, Adams read a petition from nine ladies of Virginia asking for slavery to be abolished.

The House ignored the petition and moved on.

Congressman John Patton of Virginia asked how southern women could have sent an abolition petition.

After researching it, he reported back to Congress that the petition was from southern women. They were free black ladies of *"scandalous"* reputation. He demanded that Adams be punished for wasting the time of Congress with this worthless petition.

Adams replied by defending the ladies' honor. How dare Patton refer to them as *"scandalous!"* Was there any evidence of his false accusation?

Adams: *"Mr. Patton says that he knows these women and claims that they are scandalous. Just how does he know that?"*

Patton: *"I don't know them personally. I heard this from others."*[241]

Adams: *"Just what could make them scandalous? Is it their skin color?"*

He paused as the Congressmen began to realize where he was going with this.

Adams: *"No! The scandal is not their color but their masters."*

Staring them down, Adams drove the knife deeper. *"I understand the South has many colored children who strongly resemble the masters who claim to own them."*[242]

That crossed the line. Adams' voice was drowned out as the other Congressmen exploded in fury. They didn't want anyone to know their dark secrets. They were deeply offended that Adams wanted them to face their evil. They fought back with a fury.

Mr. Dromgoogle of Virginia accused Adams of *"Giving color to an idea."*

Mr. Thompson threatened to indict Adams before a grand jury and bet his life that *"We shall yet see Adams in a prison cell."*[243]

Adams retorted, *"Let Mr. Thompson go home and study the principles of civil liberty. If this is the opinion of the slaveholding Congressmen on this floor then I want the country to know who they are. Let everyone hear this debate and we will see what the people of this nation think."*

Thompson tried to defend himself by explaining that he was only referring to the laws of South Carolina that anyone presenting the state legislature with a petition from slaves would be jailed.

Adams: *"Then thank God I am not a citizen of South Carolina!"*[244]

The room exploded in anger. *"Great sensation"* erupted in the House as the other Congressmen screamed back in fury.

This time the uproar in the House was so violent that the official Congressional record struggled to describe it with the words *"great sensation."*

Adams (raising his voice): *"Do you really think that all your threats can frighten me? Your anger and all the grand juries in the universe will never stop me from fulfilling my purpose."*[245]

When the noise died down, Mr. Rayner from North Carolina sarcastically asked Adams if he would present a petition asking Congress to abolish the institution of marriage.

Adams: *"The most damning sin of slavery is that it does abolish the institution of marriage!"*[246]

More shouts and threats interrupted Adams. Having made his point he sat back down while taunting them, *"If you don't want answers, then don't ask questions."*

Rayner replied by threatening the abolitionists, *"Long before you accomplish your purpose, every stream will run red with your blood. We will trample you under our feet."*[247]

To stop this from happening again, Congress passed a resolution *"That the right of petition does not belong to slaves. No petition from them can be presented to this House without diminishing the rights of the slaveholding states and endangering the Union."*[248]

Adams fought back with a furious speech about how denying civil rights to slaves would destroy civil rights for everyone in the nation.

He demanded to know why the House kept trying to bury the issue. *"Why won't you discuss this question? You cannot keep it out of Congress. If you persist, before you know it you will have a civil war raging while you are excluding the only way to prevent it."*[249]

Turning to face the men that hated him, Adams concluded, *"God is my judge. I have done my duty and I will do it again tomorrow."*[250]

Year after year, Adams waged war in Congress. The harder he fought, the more they fought back. They hated him.

Describing it in his diary he wrote, *"The opposition in Congress has increased way beyond what I expected."*

"One hundred members of the House represent slaves; most of whom would crucify me if their votes could erect the cross. Forty members from the free states, yet united with slavery and fake democracy, would break me on the wheel, if their votes could turn it round."[251]

"Most of rest are either so cold or so lukewarm that they are ready to desert me at my very first symptom of mistake. To be forsaken by everyone seems to be the destiny of my last days. What sustains me is an overruling awareness of justice."

As the pressure grew against him, Adams struggled to stay on the right path. Being the leading abolitionist in America was a very tough job. One mistake could cost him dearly.

Recognizing the danger, he carefully planned every move with lots of time and effort. Trying to move forward as fast as he could, he also tried not to go too far.

Describing the pressure he felt, he wrote, *"It feels like every step I take is on the edge of a cliff. Both political parties watch me, hoping I will make a mistake that they can use to destroy me in the press."*

The rest of the abolitionists weren't helping him. More divided than united, the abolitionists fought amongst themselves over how exactly they should promote abolition. Too many of them didn't understand politics or the art of fighting a public relations war.

This was very frustrating to Adams who complained in his diary. *"The abolitionists constantly urge me to do rude things, which would ruin me and weaken the cause. Then my own family pressures me to avoid any connection with the abolitionists as public opinion fluctuates on the question."*[252]

Adams fought this battle on his knees in prayer. Not until after Adam's death, when his private diary was published, did the nation learn how often he had fasted and prayed that God would give him just the right words to say.

One time he wrote in his journal, *"My mind is totally absorbed by fighting for the liberties of my country. I'm in the midst of a fiery ordeal."*

"Day and night are consumed in struggling to avoid my ruin! God, send me a good deliverance!"

Praying fervently, Adams cried out to God for guidance on how to stay on the right path without falling into the error of either foolishly rushing *"into the fiery furnace"* or the *"cowardly spirit of shrinking from the danger of my duty."*

His diary also documented that he frequently attended church and the sermons often encouraged him. Like the time the pastor preached on Philippians 1:6 that God would complete the good work which had been started.[253]

While the battle continued, time passed. Congress adjourned and a new President was elected.

On March 4, 1837, President Martin Van Buren was sworn into office. He promised to veto any bills from Congress that interfered with slavery.

That prompted Adams to introduce defiant legislation. He authored a bill in Congress to abolish slavery.

That bill was laughed at and promptly dismissed.

Adams fought back with another bill to eliminate the 3/5ths rule, which would destroy the proslavery majority in Congress.

That bill was also soundly defeated. Adams watched this happen with a smirk on his face. He knew he was winning in public opinion even while losing in Congress.

The next thing he went after was the *"gag rule."*

Ever since Congress had tried to silence all discussion of slavery, Adams had found every possible way of forcing them to keep facing the issue.

The more they opposed him, the harder he fought. That's how he kept the antislavery movement on the front page of the newspaper.

On Friday, January 21, 1842, Adams decided to push the envelope even further.

Pouring over the thousands of petitions received by Congress, Adams selected some that would be considered most offensive.

When it was his time to have the floor, Adams read the petition *"From a number of Massachusetts citizens asking that the naturalization laws may be changed to permit free colored immigrants to become citizens of the United States and be able to hold real estate."*

Mr. Wise jumped up and *"Raised the question whether that petition could be received."*

The House voted and it was tabled.

Mr. Adams reached back into his desk and presented another petition from voters in Massachusetts.

This petition reminded everyone that the Constitution guaranteed a form of government to each state that would protect the rights of the people. *"Yet there are thirteen slaveholding states whose governments are absolutely tyrannical, troublesome, and oppressive in hurting a great number of its citizens. Congress must act to adopt some feasible measures by which this alarming evil may be corrected."*

Adams couldn't get through more than a few sentences before being interrupted again.

Mr. Jones of Maryland rose and demanded the petition be laid on the table.

Adams was nowhere near finished. Pulling out the next piece of paper, he read a petition from the Pennsylvania Antislavery Society.

This petition was concerned that America was being forced into a war with England, *"For the purpose of forcing the British government to assist in holding natives of the United States in slavery. Such war would exceed in unrighteousness that which was waged against this country by England in 1776 as the wrongs inflicted on the slaves greatly exceed the wrongs which led to the Declaration of Independence."*

(This was a real issue. After England abolished slavery, the British Navy patrolled the ocean, stopping and searching other nation's slave ships. Trying to escape them, slave traders hoisted the American flag and tried to hide behind the Fourth Amendment forbidding search and seizure. That didn't stop England from targeting slave traders. They disrupted the slave trade on the high seas, resulting in a bitter battle in Congress over whether to declare war on Great Britain for interfering with American shipping. For further details, read *The Navy and the Slave Trade* by Christopher Lloyd.)

That was all Adams could say before his voice was again drowned out by shouts and curses from the other Congressmen present. Rage poured forth as they demanded that Adams be silenced. They couldn't believe he would dare talk about something so offensive to them.

The Speaker of the House demanded that Adams stop reading the petition.

Adams kept going. The other Congressmen continued to interrupt him. The intensity of the moment was described in the official Congressional record. *"Mr. Adams was frequently called to order by Mr. Wise and declared to be out of order by the Speaker."*

"But after each interruption he continued reading the paper until he had gotten through the whole of it...(Congress was filled).....with much noise and excitement prevailing at the time."[254]

Across the room Congressman Joshua Giddings of Ohio (who was a loyal friend to Adams) watched the *"Deep hostility as the slaveholders and northern Democrats became greatly excited. Disorder increased amidst shouts of protest, insults, and threats. This storm of passion raged with more vindictive feeling than ever before."*[255]

Sitting next to Adams was his personal assistant and legal researcher, Theodore Weld, who described the uproar in a letter to his wife. *"The day before, Mr. Adams had said that he should present some petitions that would set them in a blaze. So I made sure to be in the House of Representatives at the time."*

"I've never seen such a scene, as the old man fought the storm and dealt his blows upon the head of the monster, lifting up his voice like a trumpet till slaveholders absolutely writhed and howled under his dissecting knife."

Weld was impressed by how Adams could talk *"With as much energy and zeal as a Methodist at a camp meeting."*[256]

Adams was called to order by the Speaker of the House.

Speaker: *"The gentleman from Massachusetts must not read the entire petition without permission of the House. Limit yourself to a brief summary of its contents."*

Adams: *"Well, Sir, I am giving a brief statement of its contents."*

Mr. Wise: *"The question is whether the petition is presentable at all. It is not a petition but a series of resolutions."*

Adams: *"It's not a series of resolutions, but a single resolution."*

The Speaker of the House had the clerk read the rules of debate in the House.

Adams proposed that the paper be printed in the Congressional Journal.

Speaker of the House refused.

Adams requested that an entry should be made that he had requested the printing of this paper.

Mr. Weller interjected: *"That can't be done. It wasn't read."*

Adams: *"You may hear more of this subject before the session is over."*

Moving right along, Adams pulled out another petition from the citizens of Massachusetts declaring that if there was a slave revolt in the country they would REFUSE to *"Take up arms to protect slaveholders."*[257]

Then Adams turned to stare down the powerful Speaker of the House and future President James Polk. Furious at how *"This House stifles debate,"* Adams defiantly declared, *"How intensely we are forced to feel that the Speaker of this House is a slaveholder himself."*

Adams face tightened with righteous indignation as he continued, *"I see where the shoe pinches, Mr. Speaker, and it will pinch more yet."*

"I'll deal out to the Congressmen a diet that they'll find hard to swallow. Before I'm finished, every slaveholder, slave trader and slave breeder on this floor will be ashamed of himself."[258]

Speaker of the House (raised his voice): *"Mr. Adams is out of order and must take his seat."*

Adams: *"I'm in my seat!"*

Then Adams, still holding the floor, presented another petition. This one came from *"Forty-one citizens, who are colored seaman of the United States, saying that upon visiting some Southern ports, they were imprisoned in violation of the Constitution without being accused of any crime but their color."*[259]

Mr. Weller rose and moved to lay the petition on the table.

Adams continued, *"I have another petition to present, which unfortunately is deeply personal to myself. Some respectable citizens of Georgia complain that I am chairman of the Committee on Foreign Relations and demand that the House remove me. I demand the right to defend myself against this."*

Mr. Wise rose to object.

Adams: *"How can you object, when you are one of the people attempting to remove me from this committee?"*

Mr. Habersham rose to claim that the petition was a hoax. He knew the citizens of that county in Georgia and not one of the names resided there.

Speaker of the House asked that the petition be read in full.

Adams read from the paper that these citizens believe that he *"Is possessed by mania on all subjects connected with people as dark as a Mexican and therefore he is not fit to be entrusted with the business of our relations with Mexico."*

Mr. Wise moved to lay it all on the table and move on.

House voted and refused.

Mr. Underwood moved to adjourn.

House voted to continue.

Mr. Marshall of Kentucky asked Adams to prove that he is not *"A maniac on a particular subject."*

House tabled that and adjourned.

House reconvened and Adams demanded the right to defend his position on the committee. *"One hundred questions of order were raised trying to gag me. This charge is not a trifling one. Gentleman in the House believe I am laboring under mania."*

Mr. Wise: *"It's true."*

Adams: *"Then if it's true, the House should consider it and find someone to replace me as chairman. But until the fact is proved, allow me the privilege allowed to all suspects of being considered innocent."*

"The entire slave-trading representation of the House is against me and even the Northern partners of the slave traders give them a majority. There is an alliance between the Southern slave traders and Northern Democrats. I am presenting evidence that the South would punish all the Northern Congressmen who are opposed to slave trading."

That brought a firestorm of interruptions. They couldn't believe Adams had just pulled the cover off the backdoor deals controlling politics.

Speaker: *"The gentleman from Massachusetts must remember that he is only speaking by the permission of the House to defend himself and must confine himself to reasonable limitations."*

Adams: *"Why didn't the Speaker confine the gentleman from South Carolina to the question?"*

"I thank the gentleman from South Carolina for throwing open the whole subject. My defense is proving that some Congressmen want to get me kicked off the committee. I have a letter to read to prove it."

Speaker: *"You can't read a letter accusing someone not present to defend themselves."*

Adams: *"The gentleman is present."*

Speaker: *"True, but the House objects."*

Adams: *"May I have permission to read a small part of the letter?"*[260]

Speaker denied it.

Adams appealed.

House vote taken. Permission granted by 78 to 76 vote.

Adams read the letter from Mr. Wise to his constituents complaining of the antislavery men appointed to Congressional committees.

Mr. Rayner objected.

House adjourned.

House reconvened on Monday January 24, 1842.

Still having the floor, Mr. Adams rose to remind the House that he was interrupted in reading a letter from a member of the House.

Speaker of the House had the clerk read the rules.

Mr. Fillmore (who later became President in 1850) requested the House begin addressing some other pressing bills.

Mr. Wise requested the House allow Mr. Adams to proceed.

Vote taken. Adams was allowed to continue.

Adams hoped the day would come when he would have an opportunity of opening the whole subject of slavery for debate and then Mr. Wise would have the opportunity of emptying his whole heart.

Speaker of the House ordered Adams to move on to this next petition.

Adams presented several more antislavery petitions, which were all promptly rejected by the House.

Then Adams presented the most offensive petition on how the Northern citizens were tired of their tax dollars protecting slavery. The North demanded to secede from the South!

Adams: *"Forty-six citizens of Haverhill, Massachusetts ask that you will immediately adopt measures peaceably to DISSOLVE THE UNION, because most of the resources of the North are drained to sustain the views and course of the South without any adequate return. Judging from history if this pathway is persisted, it will certainly overwhelm the whole nation in utter destruction."*[261]

Adams moved that the petition be referred to a committee who would have to draft a written response showing why the request should not be granted.

This brilliant move would force the slavery caucus to justify their views, which of course they couldn't do. But the proslavery caucus had been waiting for this chance. They needed something to make an example of him. Now they had a chance to put him on trial for something not directly related to slavery. Or so they thought.

Mr. Hopkins jumped up and demanded the petition be burnt along with the man presenting it.

Mr. Wise moved to censure Adams.

Adams: *"Good!"*

Mr. Meriwether of Georgia proposed to table the petition.

Adams was shocked that *"Such an objection should come from a state who has so often considered leaving the Union."*

Mr. Chapman moved that the House adjourn.

Adams: *"If you're going to censure me, do it now."*[262]

Vote taken. House refused to adjourn by 87 to 49.

House voted again and adjourned.

The news spread quickly across America. People were shocked to open their newspapers to hear that former president Adams was to be put on trial for treason, because he had presented a petition that the North wanted to secede from the South!

That night, the proslavery congressmen met together to plot exactly how to punish Adams. How could they make an example of him that would intimidate everyone?

Plans were made. Congressman Marshall of Kentucky was selected to lead the prosecution. They figured as a handsome young politician, Marshall's popularity would help him win the nation to his side.

ADAMS IS PUT ON TRIAL IN CONGRESS

House reconvened on Tuesday January 25, 1842.

Every seat in the gallery was filled and many people standing. Newspaper reporters crowded in to cover the story making headlines across the country.

The whole nation was following along through the newspapers as the last living link to the Founding Fathers was put on trial for treason. Silence reigned throughout the room as the people waited in breathless anticipation to hear Adam defend himself.

First, the prosecution made their case. Mr. Marshall launched into a long speech accusing Adams of *"Personal bitterness and a determined spirit of hate and vengeance towards slave states."*

Marshall thundered, *"God forbid that I would draw a line to array the Southern men on one side and Northern on the other! Surely the North is bound by love of country to oppose this petition as strenuously as the South does."*

Then Marshall demanded that Adams come see slavery in the South for himself, instead of just reading about it in books.

Adams: *"You'd lynch me."*

Marshall: *"Most likely. But then you would be persuaded to depart from your path."*

Adams: *"Not an inch!"*

Marshall said he didn't want to get into a discussion about the grievances of slavery.

Adams: *"You'd better not!"*[263]

Marshall continued on with the proposed resolution to punish him. *"The dissolution of the Union implies the extinction of our nation."*

"Thus, Adams has offered the deepest insult to the people of the United States by the crime of high treason. If this outrage be permitted to pass unpunished, he will have disgraced his country in the view of the whole world."

"The House deems it an act of grace and mercy when they only punish him with their severest censure. For the rest they turn him over to his own conscience and the anger of all true American citizens."[264]

Marshall finished, *"How ironic that the man igniting the fire to destroy our liberty is the son of John Adams who laid the cornerstone of our freedom."*[265]

Adams was finally given a chance to reply. He started by discussing the Southern friends he had once had: George Washington, Thomas Jefferson, James Monroe, and James Madison. These Founding Fathers and early Presidents had all respected him and appointed him to top diplomatic positions.

Then Adams wondered how this new generation of Southerners could attack him for treason. What did they know about constitutional law?

"Mr. Marshall, claiming to be motivated by nothing other than a love of the Union with no personal hatred towards me, has accused me of treason."

"What is treason? The Constitution of the United States defines it and it is not for Marshall..."

(Adams raised his voice) "OR HIS PUNY MIND to define it. *Where did Marshall get his legal theories? Definitely not from his uncle."*

(His uncle was John Marshall, Chief Justice of the Supreme Court. Before being appointed to the Supreme Court, John Marshall had represented the abolitionist side of a landmark court case in Virginia, *Pleasants v. Pleasants*, which had enforced provisions of a will freeing several hundred slaves.)

Adams: *"Let him go back to law school and learn a little of the citizen's rights. I desire the clerk to read the first paragraph of the Declaration of Independence!"*

Clerk fumbled with paperwork, looking for it.

Adams repeated louder:

"THE FIRST PARAGRAPH OF

THE DECLARATION OF INDEPENDENCE!"

Clerk threw documents around, trying to locate it.

Adams: <u>*"THE FIRST PARAGRAPH OF*</u>

<u>*THE DECLARATION OF INDEPENDENCE!!!!"*</u>

Clerk: *"When in the course of human events, it becomes necessary for one people to dissolve the political bands which have connected them with another, and to assume.......the separate and equal station to which the Laws of Nature and of Nature's God entitle them.........."*

Adams: *"Proceed, proceed, proceed, down to 'right and duty!'"*

Clerk: *"We hold these truths to be self-evident, that all men are created equal, that they are endowed by their Creator with certain unalienable rights—that among these are life, liberty, and the pursuit of happiness."*

"That to secure these rights, governments are instituted among men, deriving their just powers from the consent of the governed. That whenever any form of government becomes destructive of these ends, it is the Right of the People to alter or abolish it and to institute new government....."

Adams interrupted, slamming the table in front of him. *"RIGHT AND DUTY TO ALTER AND ABOLISH IT!"*

"Now sir, if there is a principle established by the document just read, it is the right of the people to alter, to change the government if it becomes oppressive to them."

Then Adams began to explain the unconstitutionality of slavery.

The purpose of government as defined by the *Declaration* was to protect the civil rights of the people.

Thus, slavery could not legally exist in America because it violated the civil rights of the people and the American principle of government by consent of the governed. Therefore it had to be destroyed, but be destroyed within the democratic process.

By Adams presenting the petition that the North wanted to secede from the South, he was utilizing the democratic process to fight slavery and thus could not be considered an act of treason.

He made it clear, he didn't want to destroy the federal government but only to continue to exercise the many options available to work within the system to end slavery by utilizing the *"Resources by which their rights might be restored without resorting to that last remedy retained in the Declaration of Independence."*

Adam's didn't want that last remedy, which would be a total separation of North and South, like the American colonies had once separated from England.

Continuing on his theme of the unconstitutionality of slavery, Adams described how it had to be destroyed because it denied the right to due process under the law. If it wasn't destroyed, then it would continue to spread across the nation until eventually all civil rights in America would be destroyed.

Adams threatened to expose the secret conspiracies going on behind closed doors to force slavery into every corner of the nation.

"I shall show that the part of the country, from which Mr. Marshall comes, is trying to destroy the right of habeas corpus and the right of trial by jury."

"I will prove that there is an organized plot to destroy all the principles of civil liberty in the free states."

"Not just to preserve slavery within their own jurisdiction but to force their detested principles of slavery into all the free states."

"I will show that they are trying to force this country into a war with England for the purpose of protecting the slave trade. It's time that the people should arouse the nation to protect our constitutional rights."

When Adams was satisfied that he had completely ripped apart the prosecution's case, he finished with one last resounding statement.

"I rest my case upon the Declaration of Independence!"[266]

Having finished his opening statements, Adams sat down.

Mr. Wise rose up to continue the prosecution. He asked the clerk to read from the farewell address of George Washington when he was leaving public office after serving two terms as President.

In this speech, Washington emphasized that all the individual states in America needed to come together and stop fighting over minor issues.

Clerk (reading the book given him earlier that day by Mr. Wise): *"The union is now precious to you. It is a main pillar of your real independence, supporting peace at home, your safety and the very liberty, which you so highly value. Watch for its preservation with jealous anxiety."*

Adams: *"You should have thought of that at the time you put on the gag."*

Mr. Wise proceeded to accuse the British of working through American abolitionists to destroy the country.

Turning to Adams, Mr. Wise snarled, *"Mr. Adams had declared that in case of a slave revolt, if the North was forced by the Constitution to suppress the rebellion, the end result would be emancipation."*

Adams: *"Yes, sir!"*

Wise: *"To think this man came to this House from the Presidential chair, as one of the greatest diplomats, one of the highest authorities on international law and the last link that bound this age to the Revolutionary Fathers. Yet he has outlived his fame! He is now dead as Arnold. Dead as Burr!"*

(Mr. Wise is accusing Adams of being like traitor Benedict Arnold and Aaron Burr who was disgraced for killing Alexander Hamilton in a duel.)

When Mr. Wise had finished, Adams was finally allowed a chance to defend himself. The gallery was packed with people waiting to see what he would say.

First he asked that the charges against him be read again. Then he called for reading the Sixth Amendment to the Constitution.

Clerk: *"In all criminal prosecutions, the accused shall enjoy the right to a speedy and public trial, by an impartial jury of the state where the crime was committed."*

Adams: *"I wish to speak respectfully to the slaveholders. One hundred Congressmen are slaveholders. Are any of them impartial? No! Every slaveholder has the most vile motive, a personal financial interest which governs him."*

"Slaveholders are not the judges by whom I ought to be judged."

"The House has already tried and found me guilty of their accusation of perjury and high treason. Yet this House has no right to try, much less punish me for this."

"If they say they will punish me. I defy them!"

"If they offer to mercifully spare me from expulsion, I scorn and cast their mercy aside!"[267]

As Adams spoke, the unusual atmosphere in the room was noticed by Rep. Joshua Giddings. *"Breathless silence reigned. There was no loud breathing, no rustling of garments. Reporters laid down their pens and slaveholders were melted to tears."*[268]

Adams still wasn't finished. He reminded the House of the day when Mr. Wise had literally walked into the chamber covered in the blood of a man he had just killed in a gentleman's duel.

Adams thundered, *"Mr. Wise walked in Congress with his hands and face dripping with the blood of murder. The question was asked—should he be tried by this House for that crime?"*

"I opposed the trial of that man by this House because I thought he should be tried where he might have the advantage of the Constitutional right to trial by an impartial jury."

Adams turned to face Mr. Wise.

Adams: *"I saved this blood stained man from the censure of the House at that time!"*

Mr. Wise jumped up to defend himself.

Adams: *"Oh now he demands the rules be followed, does he?"*[269]

Adams did not allow himself to be interrupted but proceeded on with reading an old letter written by Mr. Marshall to his local newspaper.

Reading the letter, Adams showed Marshall had once been an abolitionist. Then Marshall sacrificed his principles to advance his political career and gain a seat in Congress. The more Adams read, the more Marshall trembled in his seat.

Later that night in his diary, Adams wrote how much he enjoyed watching Marshall, *"Writhe under it in agony. Before I had read halfway through, Mr. Saunders interrupted, objecting that I had no right to discuss the subject of slavery."*[270]

By the time Adams had finished, Marshall was totally humiliated. When the session adjourned, he dashed outside as fast as he could. His friends rushed over to try to make him feel better.

They watched as Marshall turned to Mr. Campbell of South Carolina, and moaned, *"I wish I were dead."*

Campbell: *"You're too sensitive."*

Marshall: *"(Cursing) I would rather die a thousand deaths than have to face that old man again."*[271]

Adams was not finished. His defense dragged on for days.

He attacked every aspect of slavery by putting the whole institution on trial. He accused powerful people of conspiring to protect it. He subpoenaed incriminating documents, some people did not want the nation to see.

He presented examples of how the constitutional rights of African Americans were being violated. He discussed the legal basis for their civil rights.

He even read from the death threats he had received, saying, *"If they want to murder me, let them try."*

Then he reminded the country, *"What good is a seat in Congress unless you can serve your country by occupying it?"*[272]

On and on it went until Mr. Merriwether of Georgia complained that Adams had occupied ten full days. *"How much longer would he need for his defense?"*

Adams smiled and replied, *"When Sir Warren Hastings was arraigned before the British Parliament, the trial occupied several years. Mr. Burke even occupied three months in one speech. This case precedent applies to my trial but I promise to close my defense as early as possible. Probably in only NINETY DAYS."*[273]

Throughout the room, eyes rolled and Congressmen moaned in frustration. Couldn't they just move on?

Finally Congressman Fillmore rose and said the matter had gone far enough. He proposed to dismiss it and move on.

The House voted but the trial continued.

This time, Adams turned to Mr. Wise and asked what had they gained by the *"gag rule"*?

How could presenting a petition to Congress be considered a crime? *"We read in the Bible that when Michael the Archangel quarreled with the devil....."*

Adams paused as the audience started laughing.

He continued, *"Michael used no personal insults against him. These profound lawyers need to read their Bibles and learn that when they try to ruin a man's character and usefulness to the country, they better not use personal insults."*[274]

More eyes rolled in the audience as his words sank into their hearts. It was obvious that this trial had only given Adams a bigger platform.

They were never going to defeat him.

Mr. Botts of Virginia stood up and moved to end the trial. It was time to get on with other Congressional business.

This resolution passed by a vote of 106 to 93.

Adams went home that night so exhausted he was *"Barely able to crawl up to my room but with the sound of triumph ringing in my ear."*[275]

He had won the battle but the war was far from over. While he had beaten the trial for treason, the *"gag rule"* was still there. Abolition petitions were still forbidden.

The fight would continue for years.

Year after year, Adams never backed down. Standing his ground, he made Congress discuss the forbidden topic as much as possible. That infuriated the proslavery caucus, which fought back even harder.

Senator John Calhoun was an old friend of Adams. Many years before, Calhoun had served as Vice President when Adams was in the White House.

Now that Calhoun had been elected to the Senate, he had become an enemy to Adams. He tried to stop Adams by introducing legislation to punish abolitionists. When his bill came up for a vote, Calhoun gave a long speech on the floor of Congress.

He referred to Adams as *"A mischievous, bad, old man."*[276]

Then Calhoun blamed them for causing all the trouble in the nation. *"The abolitionists attack slavery because it is wicked and sinful. I wish to confront them distinctly on that point. Abolitionists are nothing more than religious fanatics."*[277]

The Senate applauded and passed resolutions condemning, *"Any state or citizens interfering with slavery on the grounds of it being immoral or sinful."*

This was a test vote preparing the way for slavery to engulf the entire country.

Senator (and future President) James Buchanan seconded the motion. He said that abolitionists were the very reason slavery had not yet been abolished because they scattered, *"Arrows, firebrands, and death."*[278]

(Buchanan was accusing abolitionists of the bad conduct described in Proverbs 26:18.)

The attacks against Adams escalated.

On the floor of Congress, Mr. Dillett of Alabama tried to use Adam's words against him.

He quoted from a speech Adams had given to a Pittsburg meeting of African Americans, saying, *"We know that the day of your redemption must come. The time and manner of its coming we know not. Whether it may come in peace or in blood, let it come."*

Sitting in his Congressional seat, Adams replied, *"I say now, let it come!!!"*

Dillett exclaimed, *"Yes, he says let it come, though it costs the blood of thousands of white men."*

Adams: *"Though it costs the blood of MILLIONS of white men, let it come!"*[279]

A deep shudder raced through the room as the Congressmen had to think about the unthinkable.

Adams roared, *"The Union will fall before slavery, or it will fall before the Union."*

"Every beat of my heart pulses with unyielding opposition against slavery. I will continue to resist until my body returns to the dust of the Earth and my spirit to God."[280]

Those words were printed in the newspapers. Some people were inspired by them. Some people were highly offended and decided to take matters into their own hands.

Death threats poured into Adam's office. Letter after letter tried to intimidate him into silence.

Adams just smiled and walked back into Congress with more petitions.

One time he read one from two hundred twenty citizens of Illinois asking Congress, *"To acknowledge the Christian religion, by securing to the people the right to enjoy, life, liberty and happiness."*

That sparked a long debate.

Adams lectured Congress on their duty to *"Do unto others as they would have others do unto them."*[281]

The other members of Congress decided to make it clear freedom would not be tolerated.

Senator Preston of South Carolina declared on the floor of Congress that any abolitionist, who dared to visit the South, would be hung from the nearest tree. As usual, his words were printed in the newspaper.

Two weeks later, Adams responded by presenting three hundred and fifty petitions to Congress. One requested *"That Congress would take measures to protect citizens of the North going to the South from the danger of being hanged."*

Of course, Adams was interrupted by deafening shouts of protest in the House.

Yet things had changed. This time Adams noticed, *"The Speaker of the House didn't dare tell me that I couldn't proceed without permission of the House; and I proceeded."*[282]

Adam's words had pierced even the hardest of hearts in Congress. After being forced to listen to Adams every day, the Speaker of the House (and future President James Polk) would do something considered unusual in that time.

Polk wrote a will to free all of his slaves upon the deaths of himself and his wife. Fortunately they wouldn't have to wait that long. The Emancipation Proclamation would free them twenty-eight years before Polk and his wife passed away.

Meanwhile, Adams fought on until he had worn out the opposition and public opinion was on his side.

It took Congress years to realize that the *"gag rule"* had only helped Adams by giving him a bigger platform for his message. Eight years after it had begun, in 1844, Congress passed Adam's motion to repeal the *"gag rule."*

That small victory greatly encouraged Adams. Yet even as he celebrated, his heart grieved for the millions of families suffering under the horrible system of slavery.

As Adams reflected on his life, he hoped that maybe the next generation would accomplish his dream. *"All I can do is open the way for others."*[283]

On April 15, 1842, Adams stood up in Congress and gave the most important speech of his life. He gave the nation the blueprint for emancipation. Making sure that everyone heard him clearly, Adams turned to face the Southern Congressmen.

He said, *"Every time we talk about slavery, people get so offended that it raises a hurricane size mass of controversy."*

"But one day civil war will come to this nation and things will change."

"When the blood of our constituents is being shed on the battlefields of civil war, then the laws of war will take precedence over the laws of slavery, and the President of the United States will have the power to order universal emancipation."[284]

As Adams talked, they listened, but they never dreamed his words would come true.

When Adams finished and sat down, Mr. Calhoun jumped up to ask Mr. Andrews of Kentucky to give a rebuttal to Adams.

Mr. Andrews ignored the request.

Everyone else sat in cold silence until Mr. Stanley moved that the House adjourn. They adjourned.

(Senator Calhoun was also warned in a dream not to push the South to secede. In his dream, George Washington warned him that civil war would hurt the country. However Calhoun ignored the warning and persuaded the South to secede to prevent the federal government from freeing the slaves.)

Adams went home and wrote in his diary that before he died, *"I want my precise opinions to be publicly known. May God grant that they contribute to the final end of slavery."*[285]

So he published pamphlets describing his thoughts on constitutional law. He knew that many people would read it. He knew they would criticize him for it. But he had no idea that his words had already reached the right person.

Somewhere in Illinois, a grocery store was going bankrupt because its owner, Abraham Lincoln, was too busy reading newspapers.

Day after day, Lincoln was learning the reasoning he would use years later to guide the nation.

Many years later, when he approved the Emancipation Proclamation, Lincoln echoed Adam's words when he said it was a *"Necessary war measure for suppressing said rebellion."*[286]

Yet even after accomplishing all this, Adams still had more work to do. While the battle was being fought in Congress, Adams had been asked to intervene in another crisis.

THE AMISTAD CASE

In 1839, Portuguese slave traders kidnapped fifty-three Mendi people from Africa and shipped them to Cuba. There the forty-nine adults and four children were sold to Spanish slave traders: Jose Ruiz and Pedro Montez. The slave traders put them on the *Amistad* and sailed for a plantation in the Caribbean.

They never made it.

At sea, the Africans found a chance to escape and broke free from their chains. The crew tried to subdue them but the Africans were stronger.

Seizing control of the ship, they killed the ship's captain, tied up Ruiz and Montez and ordered the crew to sail the ship back to Africa.

Montez was put in charge of navigating the ship. He tricked them. During the day, he turned the ship to sail towards Africa. At night he directed the ship towards America, hoping to get there before anyone noticed. For days the ship zig zagged across the ocean until it arrived in New England.

Upon approaching American waters off the coast of New York, the *Amistad* was stopped by the *SS Washington.*

The Africans were arrested and charged with the crime of mutiny.

The crew members were not arrested. Slave traders Jose Ruiz and Pedro Montez departed the ship and found a comfortable place to stay.

What came next would be one of the most unique legal battles in American history.

Slave traders Ruiz and Montez filed a lawsuit demanding that their *"property"* be returned to them, so they could return to Cuba and sell them to the highest bidder.

Since Cuba was a colony of Spain, the Spanish Ambassador showed up in court. He claimed American courts had no jurisdiction over the matter due to a treaty that American had signed with Spain in 1795. Spain wanted these slaves extradited to Cuba to stand trial for murder.

Lieutenant Thomas Gedney, the ship captain who had discovered the *Amistad* entering American waters, appealed to the court for salvage rights. He insisted that both the ship and its human cargo belonged to him as a reward for rescuing Ruiz and Montez.

Lewis Tappan wasn't going to allow it. As an abolitionist, he hated slavery. As a successful textile businessman he had the money to fight it.

Hiring lawyers, he filed a lawsuit, demanding the Africans be freed. He asked the court to punish Ruiz and Montez for kidnapping and false imprisonment.

That was not good for the proslavery power.

Trying to sweep everything under the rug, President Martin Van Buren got involved, sending his Secretary of State, Mr. Forsyth, to resolve the matter quietly.

Mr. Forsyth ordered the local District Attorney assigned to the case, *"Be careful that no court proceedings let the slaves get beyond the control of the President."*[287]

Knowing the court would rule in his favor, President Van Buren sent a ship to sit in the harbor, ready to remove the Africans back to Cuba as soon as the court ruled.

That infuriated Lewis Tappan, who chartered his own ship and let it sit right next to the slave ship, with instructions to be ready take the Africans wherever they want to go—somewhere they would be free.

The case was in the hands of District Judge Andrew Judson. Everyone thought he was proslavery because he had already issued a court ruling shutting down a black school, but a big surprise was coming.

This judge ruled the Africans should be freed, since importing slaves from Africa to America was illegal, and the slave traders, Ruiz and Montes, had tried to cover up their crimes by forging their documents.

Both Ruiz and Montes were thrown in jail. They posted bail and fled to Cuba.

No one was prepared for that court ruling. The District Attorney quickly appealed it while President Van Buren's office scrambled to keep this scandal from growing. They didn't want this to turn the American public against slavery.

The case was argued all the way to the Supreme Court.

Lewis Tappan was thrilled at this rare opportunity to get a freedom ruling. They just needed the right lawyer who could convince the Supreme Court that slavery was unconstitutional.

Who would have the experience and reputation that the judges would listen to? With dozens of lives hanging in the balance, Tappan begged Adams to represent the Africans before the Supreme Court.

Adams was nervous about it. Suffering from an eye inflammation, he didn't feel well enough to begin the long hours of legal research necessary to prepare a proper defense.

Plus, the odds were stacked against him. This was a David vs. Goliath type battle that no one else wanted.

Most of the Supreme Court Justices owned slaves themselves. The Chief Justice was Roger Taney who was only a few years away from writing the horrific *Dred Scott* decision. How could Adams possibly convince them to change their minds?

Scribbling in his diary Adams thought: *"The world, the flesh and all the devils in hell are arrayed against any man in America who shall dare to join the standard of Almighty God to put down the African slave trade. No one else will undertake it."*

"Here I am on the verge of my seventy-fourth birthday with a shaking hand, a darkening eye, a drowsy brain and with all my faculties dropping from me one by one, as the teeth are dropping from my head. Yet my conscience presses me on. Let me die upon the breach."[288]

Drawing upon what little energy he had left, Adams began planning his legal strategies. He researched the facts, subpoenaed records, wrote extensive notes and then showed up for court with a hope and prayer. Maybe he could do the impossible.

In early 1841, the Supreme Court convened. The prosecution opened its case, demanding that the Africans be returned to Cuba to stand trial as murderers.

Then Adam's turn came. For almost five hours straight, he gave one of the most powerful defenses in legal history.

He opened by turning towards the framed copy of the *Declaration of Independence* hanging on the wall.

That was the document that had birthed America. That thousands of Americans had sacrificed their lives for in the Revolutionary War.

Pointing at it, Adams roared, *"I know of no law, except that law, which is forever before the eyes of this court."*

"No other law applies to my clients but the law of nature and of nature's God on which our fathers placed our national existence."[289]

Then Adams opened his briefcase and pulled out stacks of records showing a secret conspiracy.

Having subpoenaed all the correspondence between the Secretary of State Forsyth and the Spanish Ambassador, it was obvious what was going on behind closed doors. People in the highest levels of government were willing to bend the laws to line their own pockets.

Reading from document after document, Adams pointed the finger right at President Van Buren for allowing slaveholders to control the federal government. *"Why would the Secretary of State tell the District Attorney to keep the Africans under the control of the President?"*

"These Africans were at peace with the United States. They were not pirates. They were in possession of the ship and on a voyage back home when Lt. Gedney seized them without any warrant or authority. I ask in the name of justice, by what law was this done?"

Then Adams read from American maritime law that in case of an emergency on the high seas, *"If a ship engaged in a lawful voyage is forced by weather or other unavoidable cause into the port, then persons on board would be entitled to the same protection which the laws of New York provide for everyone within its limits."*

Adams slammed Van Buren for secretly ordering the Africans back to slavery, declaring, *"Was this order given in a country where human rights are words without meaning? It was null and void, given in the land of the Declaration of Independence!"*[290]

"The government's actions had been founded on the wrong assumption that the two Spanish slave traders were the only parties harmed. That all the right was on their side and all the wrong on the side of their surviving self-emancipated victims."

"I ask your honors was this justice? No! It was sympathy for the slave traders as stated by Secretary Forsyth himself."

"This is a court of justice where both sides have a right to be heard. I call upon this court to restrain itself in the sacred name of justice."

Launching into a history lesson, Adams described how for thousands of years many nations had considered prisoners of war as property to be bought and sold.

The supposed justification for slavery had been the theory of conquest: one nation felt they had the right to take the property of another nation if they conquered them.

Yet all this had changed among Christian nations. They had led the way in demanding that the rule of law follow the doctrine of Christ who commanded us to treat other people the way that we would want to be treated.

Describing how these principles had been written into the *Declaration of Independence*, making civil rights more important in America than the world's traditions, Adams continued.

"*Slavery has its origin in violence as the world once agreed that it was legitimate. But throughout Christian nations this harsh rule has been exploded and war is no longer considered as giving a right to enslave captives.*"

"*The Treaty of Ghent, (which Adams had negotiated) signed between England and America in 1814, declared that slavery was irreconcilable with the principles of humanity and justice. That both nations would use their best efforts to abolish it.*"

"*Then the following year, in 1815, ambassadors from Vienna, Austria, France, Prussia, England, Portugal, Russia, and Sweden issued a declaration that they would use their all their means to bring universal emancipation.*"

Then Adams played his power card. Guess who had been involved in negotiating the treaty with Spain that Spain was now invoking to demand the return of the Africans?

Adams testified from personal experience that years ago when he had negotiated that treaty with Spain, neither himself nor the Spanish Ambassador had ever intended it to protect slavery.

Next Adams moved on to "*Show that slave traders Ruiz and Montes were acting in a way forbidden by the laws of Spain and the U.S.*"

"*On March 2, 1807, the importation of slaves into America was forbidden. The law stated that if any slave ship was found within the jurisdictional limits of the United States, then that vessel could be seized and prosecuted by the U.S. Courts.*"

"*On September 23, 1817, the King of Spain said that it shall not be lawful for any subjects of Spain to purchase slaves or carry on the slave trade in any manner whatsoever.*"[291]

On and on it went as Adam's legal research shredded every piece of the prosecution's argument.

Finally he closed by asking the judges to think about eternity. One day, each one of them would have to stand before God in the courtroom of eternity, to answer for everything they had done in their lives. How would God judge them on the decision that they were going to make in this case?

Deeply moved, Justice Joseph Story, who wrote the court's majority decision, described Adam's argument as *"Extraordinary for its power and bitter sarcasm."*[292]

The Supreme Court ruled 7 to 1 that the Africans, *"Were born free and had been unlawfully kidnapped. They ought to be free."*[293]

Even more surprising, Chief Justice Roger Taney, who later wrote the horrific *Dred Scott* decision, had been persuaded by Adams.

The court ordered that the Africans were to be freed immediately. However, the *Amistad* itself was turned over to Lt. Gedney for salvage.

Tappan paid for a ship to transport the Africans wherever they wanted to go. They requested to return to their families. Shortly thereafter they sailed for Africa with several missionaries on board. Then they would disappear from the historical record.

Meanwhile, Adams had so destroyed the political career of President Martin Van Buren that he would never be elected again. Van Buren would lose both of his next two attempts to run for President.

Spain was not happy about this court decision. They demanded full payment for the loss of the *Amistad* and its human cargo which could have been sold for thousands of dollars.

Year after year, Spain lobbied Congress for the money. Proslavery Congressmen like Mr. Ingersoll of Pennsylvania introduced bills to pay Spain as much as $70,000 for the *Amistad.* Adams fought these bills with a fury until each one was crushed.

Several years later, Adams suffered a severe stroke. He was bedridden for several weeks.

With Adams gone from Congress, Senator Buchanan saw his chance and introduced a bill to pay $50,000 for the *Amistad.*

When Adams heard that it was coming up for a vote, he dragged his frail body out of bed and showed up in Congress just in time to declare *"God forbid that any claim should be allowed!"294*

Congress listened. The bill was defeated. No money would ever be paid.

Even after all that, Adams still accomplished many other remarkable things. His wisdom would intervene again in another situation.

For years Adams had advocated the land rights of Native Americans. He had even warned Congress that God's judgment would come if America didn't start treating the Native Americans better. And both Northern and Southern Congressmen considered Adams as one of the greatest experts on property law.

On February 9, 1846, Congress debated whether America or England had proper claim to the Oregon territory.

There was a treaty between them with ambiguous political jargon leaving both nations on the verge of fighting for it.

England had already spent many years disputing America's claim. With this treaty coming up for consideration, Adams was asked for his legal opinion. He stood up in Congress and gave a fascinating lesson on how the Bible is America's foundation for real estate law.

According to the official Congressional record Adams began with *"The foundation of property title is found in Genesis 1:26-28."*

The Congressional clerk read Genesis 1:26-28 (KJV):

"And God said, Let us make man in our image, after our likeness: and let them have dominion over the fish of the sea, and over the fowl of the air, and over the cattle, and over all the earth, and over every creeping thing that creeps upon the earth."

"So God created man in his own image....male and female....And God blessed them and God said to them, 'Be fruitful, and multiply, and replenish the earth, and subdue it: and have dominion over the fish of the sea, and over the fowl of the air, and over every living thing that moves upon the earth.'"

Adams: *"That is the foundation of all human territorial rights, whether of universal domain, national jurisdiction or individual property, from the right of the U.S. to the soil upon which the Capital stands, to the chair occupied by the Speaker of the House."*

"Now I will ask the Clerk to read Psalm 2:8."

Clerk: *"Ask of me, and I shall give you the heathen for your inheritance and the uttermost parts of the earth for thy possession."*

Adams: *"Our Lord Jesus Christ, after rising from the dead, said to His disciples, 'Go forth and preach to all nations my Gospel and I will be with you to the ends of the world.'"*[295]

"Now the general authority given to man to increase, multiply, replenish the Earth and subdue it, was a grant from the Creator to every individual of the human race."

"The whole human race was to decide among themselves what should be the boundaries of that portion of the Earth given individuals by the general grant from the Creator."

"When communities were formed, it became a matter of legislation among them to decide to whom any particular lot of land on which to build a house should belong. Any territorial right as between individuals was to be regulated by legislation, just as it was to be regulated by consent between nations."

"Our title to Oregon stands on the same foundation. Yet when Columbus discovered this continent, a great controversy came among nations."

"Even today this dispute has never been settled. Property titles were claimed from agreement, from conquest, treaty, discovery and exploration, but that is not the foundation of any of our titles."

"Discovery does not grant title of itself. Neither does exploration."

Then Adams read from the treaty between England and America. How it only provided that the territory *"With its harbors, bays, creeks, and rivers shall be open for the navigation of both parties without either party claiming exclusive jurisdiction during that time. That was all."*

Adams noted, *"England herself admits she has no title there. She claims Oregon for the purpose of leaving it uncultivated to benefit the Hudson Bay Company. Now she knows that it would have no value to her at all from the day that it is settled by plowers of the ground."*

"That's the difference between her claims and our claims. We claim that country for what? To make the wilderness blossom as the rose, to establish laws, to increase, multiply and subdue the earth, which we are commanded to do by the first order of God Almighty."

"That is what we claim it for. She claims to keep it open for navigation, for her hunters to hunt wild animals and of course she claims for the benefit of the wild animals. That is the difference between our claims."[296]

The Congressmen were fascinated by this impromptu Bible study.

When Adams finished and sat down, they asked him to keep going.

In the end, they followed his advice. America asserted its claim to Oregon and England let go.

Meanwhile, Adams labored on in Congress, as his body was wearing out. Knowing the end was near, he changed the title of his diary to *"Post humorous Memoirs!"*

Adams had one last wish. To never retire. He hoped to die one day while working at his desk in Congress.

That would happen on February 21, 1848. While Congress was in session, Adams suddenly collapsed over his desk.

A cerebral hemorrhage had exploded in his brain. The shout was heard in the House, *"Mr. Adams is dying!"*[297]

Everything stopped.

The Congressmen rushed over to help Mr. Adams but it was too late. His time had finally run out. Both the Senate and House adjourned as Adam's wife was summoned.

At eighty years old, Adams' last words were *"I am content."*[298]

Watching him from across the room was a young Illinois lawyer who had just been elected to Congress. Abraham Lincoln served his first term in the House of Representatives as Adams served his last.

As Adams' frail body was carried out and laid to rest, the torch of freedom passed to another man who would also lay down his life for it.

The funeral of Adams was held in the place where his heart was. The place he had labored for so many years. The room where he had fought so hard for his country: the House of Representatives.

There all of the other Congressmen came to honor someone who had had a profound effect on their lives.

One of the pallbearers serving at the funeral was that young Congressman named Abraham Lincoln. He listened intently as all the other Congressmen paid their respects to the memory of John Quincy Adams.

Mr. McDowell of Virginia described how they had all looked up to Adams for being *"Born in the day of our revolution and brought up with the founders of the Republic."*

"He was a living bond of connection between the present and the past. The lesson taught by his life is, 'Be ye also ready for ye know not the hour when Jesus comes.'"[299]

11

Theodore Weld and The Lane Rebels

*"If I were still trying to please men,
I would not be a servant of Christ."*
Galatians 1:10b (BSB)

Theodore Weld

1803-1895

From Connecticut

Theodore Weld had gotten used to violence. It seemed to follow him everywhere he went. Once upon a time he had been a college student dreaming of changing the world.

Now he was a preacher with a price on his head and angry mobs trying to disrupt every meeting he held. As the violence escalated, he knew that he was fulfilling the call of God on his life. There was no other place he wanted to be.

Weld had always been the type of person to go against the flow. When he was only six years old, attending school in the first grade, he was upset when he saw a black classmate named Jerry being bullied.

Weld stood up for him, moving his seat right next to Jerry's and daring the bullies to do anything about it. As Weld described, *"I sat with Jerry, studied and played with him. Ever since I became an abolitionist."*[300]

Like many pastor's kids, Weld grew up in the church, then rebelled. Through the prayers of his family, he came back to Christ.

The Holy Spirit dealt with his heart until he yielded to the call of God on his life. Weld trained for the ministry by serving under Evangelist Charles Finney. At that time a major revival was sweeping America. Finney was leading it. Hearts were opening to the things of God as Finney preached the gospel.

Helping him was a team of young men. One of them was Theodore Weld. While Weld was learning some things from Finney he was also learning to disagree with him. Weld stirred up major controversy by allowing the ladies to speak in the revival services. That was considered unthinkable at the time. Didn't Weld know women were supposed to be silent in church?

Finney didn't like Weld's methods until he realized that they were helping his ministry. Eventually he caved in, allowing women greater involvement than they had ever had before.

After months of working with Finney, Weld felt the Holy Spirit guiding him in a new direction.

An old friend and mentor, Major Charles Stuart, offered to pay Weld's tuition if he would return to college. Major Stuart was a retired British military officer who had once been the principal of the high school that Weld had attended.

Recognizing that Weld had a gift for public speaking, Stuart urged him to work on his skills so that he could do great things for God. Weld was very grateful for the opportunity. So when Stuart asked him for a favor he quickly agreed.

Stuart's heart was in the abolition movement. His retirement had been spent traveling around, passing out tracts, telling anyone who would listen that slavery was wrong.

In May 1831, Stuart wrote to Weld about how he was: *"Traveling the country, holding meetings wherever I can, endeavoring to awaken the conscience of the nation."*[301]

Sowing the seeds of revival, Stuart saw enough results that the British abolitionist groups would honor his accomplishments. Yet each year Stuart was getting older.

Soon he would be going on to his reward in Heaven. It was time to pass the mantle to the next generation. So he encouraged Weld to go out and preach across America.

Weld did. In between his classes, he traveled and spoke in different cities. At the time, the western frontier was rapidly expanding as families migrated across the country. New towns were springing up all over the west, many of which were too small to have a full time pastor. They appreciated ministers who would come and preach for them.

When they heard Weld, they were very impressed. Soon invitations to minister were pouring in from around the country.

In New York City, a wealthy textile merchant, Lewis Tappan, heard about Weld and invited him to speak.

Weld declined. While he was comfortable ministering to small groups on the western frontier, he was nervous about preaching to a large crowd in a big city. What if he made a mistake?

Besides he was still a student. What did he know about theology? Reading that letter made Tappan quickly write back to say, *"There is too much theology in the church now and too little gospel."*[302]

Weld liked that response. He came and preached in New York City. The Holy Spirit moved. Hearts were touched. Seeing Weld's gift for preaching, Tappan asked him to stay and pastor the new church he was opening.

Tappan had a heart for reaching everyone and wanted to open a church with free seats. This was unusual in a time when most churches rented pews to pay the bills.

Plus, it would be highly controversial by welcoming both blacks and whites and worshipping with a racially integrated choir. This was the opportunity for Weld to enjoy the comfort of a consistent salary and home pulpit.

Weld loved the idea but knew he was not called to live in New York City. Declining the job, Weld wrote to Tappan: *"You and brother Finney think I should settle in a city! No! God has marked out for me for the highways and hedges of the west!"*[303]

However, Weld recommended Finney for the position. Tappan wrote to Finney, inviting him to leave the traveling revival circuit to come and pastor. Finney accepted the invitation and moved to New York.

As Weld went back to the west, he received another invitation to pastor a church. This time it was in New Orleans. Once again he declined.

Feeling that the Holy Spirit had a different plan for him, he wanted to be ready. Soon the new opportunity arrived. It came from another invitation from Lewis Tappan.

Tappan was frustrated by how the abolition movement didn't appear to be making any progress in America. While early groups had succeeded in abolishing slavery in the northern states, it seemed like the movement had faded.

Apathy was growing. People were getting too busy raising their own families to make the sacrifices necessary for social justice. Something had to be done before the entire abolition movement died out.

The solution came to Tappan one day as he was reading a biography of William Wilberforce.

Hearing the history of how God could use one person, Tappan was inspired to form the American Anti-Slavery Society in New York.

While this offended many of Tappan's business customers, he was willing to sacrifice everything for the cause of freedom, trusting God to work the miracle. He just needed the right people to help him. One of the first people he hired was Theodore Weld.

Tappan gave Weld a salary to travel and preach across America. As Weld did, he found another opportunity.

LANE SEMINARY

While Weld was traveling through Ohio, he heard that the Lane family of Cincinnati wanted to build a new seminary to train pastors for the western frontier. The idea was to get the Baptists and Presbyterians to work together to build the seminary.

That didn't happen. The Baptists declined but the Presbyterians accepted. So the Lane family organized a committee of twenty-three trustees to develop the school. Then they donated $4,000 and gave the project over to trustees to manage.

The first question was where to build it. Just as they begin working on it, Elnathan Kemper offered to donate sixty acres.

As the son of the first Presbyterian minister who had moved to the western frontier, Elnathan had inherited a large farm from his father.

Now he wanted to see his property be used for the Lord. Elnathan said, *"I pray that this donation will advance the Kingdom of our Lord and Savior. May you always be under the influence of His Spirit. Not selfishness or prejudice."*[304]

The trustees were thrilled. Yet money was still needed for buildings, supplies and instructors. So they launched a fundraising campaign.

It failed miserably. People didn't feel comfortable giving to the seminary unless it was under the leadership of a big name minister. Running out of options, the trustees went to Lewis Tappan and asked for help.

Tappan replied that he was also skeptical of donating. How did he know the seminary would survive? Besides, weren't there already plenty of other seminaries?

Just when he was about to turn them down, Tappan received a letter from Theodore Weld that changed his mind.

While working for Finney's ministry, Weld had met a lot of young men from the South who felt called by God to enter the ministry.

They were looking for a seminary to attend. While there were a lot of seminaries already built in America, many were under the control of major church denominations, which had made slavery a forbidden topic. Wealthy donors pulled strings at the top of major church denominations to keep this issue under control.

That gave Weld an idea. Weld told Tappan that maybe this new seminary would be the perfect place to start a public debate about slavery.

Weld promised Tappan that if he would fund the seminary, Weld would recruit enough students to make it successful. Then Weld would figure out a way to discuss the issue of abolition with southern students.

Maybe some of them would change their minds while they were away from their families and open to new ideas in the collegiate atmosphere.

Tappan loved the idea although he was skeptical of its success. These potential students had come from generations of slaveholders. Their tuition would be paid by it. Their future depended on it. One day they would inherit the plantation itself. How could Weld possibly convince them to walk away from everything?

Weld already had. While he had been traveling and preaching, he had made friends with slaveholders. Many of them invited him to stay overnight at their plantations. Weld accepted their hospitality, yet still had the courage to gently confront them on their sin. One of these life changing conversations had happened in Alabama.

John Allen had invited Weld to stay at his home and meet his son William Allen. Weld stayed for a whole week, meeting the fifteen slaves of John Allen, including one who was a Baptist pastor that ministered on Sundays to slaves from several different plantations.

One night, they had dinner with a neighbor and fellow slaveholder James Birney.

At the table, Birney asked Weld, *"I've heard that you and John have been discussing the subject of slavery for a week. Tell me some of your points."*

Weld gently summarized both sides of the issue. Then he asked, *"What gives you the right to hold people as slaves?"*

Birney was impressed by Weld's *"dignity and courtesy."* But he couldn't answer the question. The more he thought about it, the more it bothered him.

Birney went home and freed his slaves. Then he became a powerful abolitionist leader, launching a newspaper that reached many people and was widely read throughout the South.

So many slaveholders were converted into abolitionists by reading it that an angry mob was sent to destroy Birney's printing press. They threw his papers and equipment into a river.

Birney bought new equipment and kept going, knowing his work was making a difference.

Later on, his son would become a Union officer in the Civil War. His son would use unconventional tactics. Whenever he needed more troops, he would find the nearest plantation and recruit slaves right off the plantation into the military.

Meanwhile, James Birney never forgot Weld. He wrote, *"I have never seen a man with the rare combination of intellectual power and Christian simplicity as him. He shared profound ideas with clear and striking illustrations."*[305]

With that intellectual power, Weld convinced Tappan to donate $20,000 to Lane Seminary on two conditions. That it would be under the leadership of an experienced and well established pastor like Lyman Beecher. And that it would go where no American college had gone before. It would be the first American college to admit black students.

When news of Tappan's generous donation spread, another $60,000 in pledges poured in from other donors who also liked having Pastor Beecher involved. They respected how he had often been successful in getting different Christian groups to come together.

They figured he would be the right person to deal with all the different religious opinions that would come to the new school.

However, Beecher himself wasn't so excited about the thought of leaving his thriving church on the east coast to move out west to Lane Seminary.

As one of the most well known and loved pastors in America at the time, Beecher was very happy with his life in Boston. He declined the job offer.

Until his friends changed his mind. They reminded him how often he had talked about moving out west. He had even said: *"The moral destiny of our nation depends on the west."*[306]

His friends insisted this was his chance to make a difference. Besides, whether the seminary succeeded or not, his personal salary would be guaranteed by Lewis Tappan. What did he have to lose?

Beecher listened to his friends and resigned from his church. In his farewell sermon he explained that he had to leave because *"Our country needs seminaries that produce preachers ministering in the power of the Holy Spirit."*

He hoped *"to inspire"* students *"to cooperate together. Be guided by wisdom from above, hold tightly to truth, and genuinely love our Lord Jesus."*[307]

Lane Seminary opened its doors in the fall of 1833. Forty students arrived to learn the Bible. One of them was Theodore Weld.

While Tappan had offered Weld a faculty job, Weld had insisted on enrolling as a student so he could focus on spreading abolitionism throughout the student body. Just like he had promised Tappan, he had succeeded in getting many of his friends to enroll.

Now he had a group that he could work on converting. He started with his best friend William Allen.

As the son of plantation owner, John Allen, William Allen had been sitting at the dinner table when Weld confronted James Birney on the sin of slavery. Watching the intense conversation unfold, Allen had secretly admired Weld's courage and wanted to be just like him.

When Lane Seminary opened its doors, Allen was excited at the chance of working closely with Weld. The more they spent time together, the more Allen changed into a hardcore abolitionist.

Then he joined Weld's secret campaign. Together they began spreading the truth around the student body. Or so they tried. The rest of the students were not ready for new ideas.

Weld got a better idea. Why not schedule an official debate on slavery? Each night for two weeks, they would discuss a different aspect of the topic.

Everyone was invited to bring their own opinion and contribute to the discussion. The debate began in February 1834. (Over twenty five years before the Civil War).

The first student to speak was William Allen. For three nights straight he asked, *"What is slavery? Before we can decide on the cure we must understand the disease."*[308]

The students were fascinated by his eyewitness description of growing up on a plantation. Not all of the students had come from the South. Some had grown up in free states. They had never seen slavery before. All they knew was the propaganda that most of the country believed at that time.

Henry Stanton was one of those students. He wrote, *"I was passing through the hall of the seminary and saw on the bulletin board that the question for debate that evening was, 'If the slaves of the South revolted, would it be the duty of the North to aid in putting it down?'"*

"I glanced at the board and never dreamed there would be more than one side to the question."

"When that evening came, it was the beginning of my life's work: fighting for forty years until the Constitution had the Thirteenth, Fourteenth and Fifteenth Amendments."[309]

Another night James Bradley spoke. As the only African American student in Lane Seminary, he had a powerful perspective to share. Born in Africa, his earliest childhood memories were of being kidnapped and sold into slavery.

Not a dry eye was left in the room as he shared the pain of being treated as property.

Yet he also shared the power of hope. No matter how many times he had been disappointed he had held onto his dream. *"I had always believed and prayed for liberty."*

"Late at night I would stay awake trying to figure out a way to make enough money to buy my freedom."

"I kept an old spelling book, working on it every chance I had until I could read easy words. I persuaded one of my young masters to teach me to write. But the second night my mistress came and scolded her son saying, 'What are you doing? If you teach him to write, he will write himself a pass and run away.'"

"That was the end of my education."

"I used to work from sunrise to sunset for my mistress. Then I would sleep four hours. Then get up and work for myself the rest of the night."

"I made collars for horses out of corn husks, sold them for fifty cents, and bought a pig. Our neighborhood had a lot of federal land. I would take my hoe to it, plant corn and fatten my hogs for sale. This way I was able to earn $300 in two years. Eventually I was able to buy my freedom for $700."

"For years I had prayed to God for an education. When I came to Cincinnati I heard of Lane Seminary. Thank God for it. Here I have been accepted and treated as a brother."

Bradley also reminded them, *"God will help those who help the oppressed. May God strengthen you in this holy cause until every wall of prejudice is broken down, the chains burst into pieces and men of every color meet at the feet of Jesus, treating each other with kindness."*

Bradley also talked about how excited he was to be a free man that could devote his life to serving God. He looked forward *"To preparing to preach the gospel."*[310]

Bradley ridiculed how the propaganda of that time said planters couldn't free the slaves because slaves were unable to take care of themselves financially.

This was ridiculous because the slaves were actually taking care of everyone on the plantation, including themselves. That made the audience laugh.

Sitting in the audience, Stanton admired how effective Bradley's speech was. *"I wish his speech could have been heard by every opponent. His brilliant logic mixed with hilarious sarcasm shredded objections by the roots for nearly an hour as the audience roared with laughter."*[311]

The debate continued. Each night they discussed a different aspect of slavery. More students came. Night after night each student took a turn sharing their opinions. Politeness and respect reigned. No one interrupted the other.

Together they sought God's guidance on the subject. Hearts were opened. The Holy Spirit touched them.

Then they took a group vote to see if their opinions had shifted.

Huntingdon Lyman from Louisiana described: *"Everyone felt that slavery was somehow wrong and to be gotten rid of. But no one was ready to call it a sin."*

"The debate was long and intense. Most of us knew little of the subject until Theodore Weld began to speak. For eighteen hours over six nights, he held the floor, giving us the origin, history, and effects of slavery."[312]

The debate went deeper. Weld had them examine proslavery propaganda pamphlets and discuss the arguments.

Their eyes opened and they fully rejected the horrible deception. Hearts changed. Shaking off generations of tradition, southern gentleman became hardcore abolitionists.

Huntingdon said, *"We are astonished at the result of our own investigation. Now we perceive that we have been deceived."*[313]

John Pierce explained: *"I was too full of prejudice to attend until the debate was almost over. Then my conscience compelled me to go and listen to two of the last nights. As Weld spoke, my blind eyes were opened. Ever since then I've been a strong abolitionist."*[314]

James Steele felt the same way. He had also entered the seminary, *"With some of the common prejudice."*

Yet things had changed. *"Just before the debate started, Porter said to me, 'Don't you realize that you will go to Heaven with colored people?'"*

"After the successful debate, prejudice never again had any place in my heart."[315]

Fellow student, James Thome, described his radical change. *"Born and raised in Kentucky, my heart had been hardened by the opinions and habits in the atmosphere of slavery."*

"Within a few months of being at Lane Seminary and the fair discussion, abolition principles seized control of my conscience. No other subject takes such strong hold of a man."

"It haunts him, drives him and rings in his ear the sounds of blood, writing 'you are the man' on the forehead of every oppressor."[316]

(Thome is referring to the story in 2Samuel 12:7 where the Prophet Nathan uses a story to confront King David on his affair with another man's wife. Nathan got King David outraged at the injustice in the scenario then told him that he was the man described.)

Thome: *"Now as the heir to a slave inheritance, I boldly denounce the whole system as an outrage, wrong, and cruelties that sicken my soul."*[317]

The students did more than just denounce slavery. They realized that true repentance required making changes.

Henry Stanton described, *"It was astonishing to watch the southern students change. One of them, Henry Thompson, had been paying his tuition with the income from hiring out slaves he owned back home in Kentucky."*

"His conscience made him go home and free them. Then he found a job. Now he's working to pay for their education."[318]

The debate ended with the students creating a new campus club. Huntingdon described their goal.

"We vowed to the Lord to use our personal efforts and whatever influence we have to raise up the free black population and to persuade our fellow men to love them as they do themselves."[319]

This campus club did not stay on campus. The students went out to the local town, sat down with the African American families, and asked what they could do to help.

Weld had already built strong relationships in the local community. Ever since coming to the seminary he had been spending time getting to know the neighbors.

Cincinnati had a large population of former slaves. Some were fugitives who could be caught and sent back at any time. Others had bought their freedom with money they made working on the side. As Weld listened to them, his heart broke with their pain. Many still worked long hours trying to earn enough money to redeem more family members.

Weld described, *"I was with the colored people by day and night. If I ate in the city it was at their tables. If I slept in the city it was at their homes."*

"If I attended parties, it was theirs: their weddings, their funerals, their religious meetings. During the eighteen months that I spent at Lane Seminary, I did not attend Dr. Beecher's church once."[320]

What the town wanted was a school, open to both children and adults, with flexible classes that people could attend at night after they worked during the day. This would be the first education opportunity many of the former slaves ever had.

So the Lane students organized a new school with classes they taught themselves. Tuition was free.

As they settled into their new schedules, they saw good things happening in the local community. People were excited about all the new doors opening to them.

Lane student Augustus Wattles noticed, *"Our colored brethren are animated with hope. Frequently, when I pass their houses, older women will stop me and ask if I think they are too old to learn."*

"When I say, 'Never!' they brighten up and reply, 'We have been slaves and have never seen things like this happen. We will come to school as soon as we finish saving enough money to redeem the rest of our family.'"[321]

The first school they opened in Cincinnati was quickly flooded with students. So they opened another. And another.

Pretty soon they had five schools and their own library but not enough teachers. Wattles wrote back east, inviting volunteers to come and help. Both the seminary students and the Tappan family opened their wallets to pay the moving expenses of the ladies who volunteered.

Yet when the ladies came to help, they found themselves harassed on the city streets. Some people did not like how they were changing the city.

Day after day, local residents publicly cursed and threatened them. As they came and went, they had to dodge filthy trash tossed at them.[322]

Cincinnati was not ready to see blacks and whites integrate as equals. Locals were shocked to see southern gentleman walking down the streets with their African American friends.

Why were these students breaking generations of tradition?

As public opinion simmered into a boil, the newspapers began denouncing the students. *Western Monthly Magazine* rebuked them and demanded that they *"Mind their own business and books."*[323]

The *Cincinnati Journal* complained in an editorial,

"These inexperienced students should stick to studying and leave social issues to others. Do they think they have the right to make rules? We will never bend to their principles!"[324]

Letters from angry citizens poured into the local newspapers.

James Hall wrote an editorial called *Education and Slavery.* He argued the subject was too complicated for students to understand. Seminaries were supposed to remain neutral. Students should be content to study and stop trying to change the world.

Weld responded with his own letter to the editor. He pointed out how students were supposed to understand the world they lived in.

"Why can't they discuss the sin of slavery when this accursed thing is exalting itself against everything of God?"

"If these students are being prepared to represent Christ, how can they refuse to think, feel and speak?"

"Aren't seminaries supposed to educate the heart as well as the head? To deepen the emotions while increasing knowledge? If not, then hire the devil to teach. He is an intellectual encyclopedia of learning!"[325]

Weld's letter to the editor sent shock waves across the city. People were talking about how Weld had made it very clear: *"The days of slavery are numbered as the nation awakes."*

Some people liked his letter so much that they mailed copies to proslavery leaders. Other people hated it so much that they threatened the newspaper.

Pressure began to build. People demanded that the seminary do something about its crazy students.

Beecher was worried. He had warned Weld not to start the debate. He knew it would disrupt the future of the seminary.

Beecher told Weld, *"If you want to teach colored schools, I can fill your pockets with money; but if you keep visiting colored families and walking with them in the streets, you will be overwhelmed. The direction you are taking will defeat your own purpose."*[326]

Beecher had different ideas for the seminary.

He wanted the students to pray more and rock the boat less. Shouldn't they wait on God's timing to change America? He told Weld, *"Pray much, say little, be humble and wait."*[327]

Even worse, Beecher actually told the students, *"Follow my example. I never take a public position on any topic until after I was fully assured that public opinion was strong enough to support me in that direction."*[328]

Weld treated Beecher with respect. He never said anything against him. Even Beecher himself admitted, *"We never quarreled."*[329]

But Weld persisted in following the Holy Spirit. Not man's control. He made it very clear to Beecher that there was no time to wait. *"These poor brethren and sisters must be helped."*[330]

When the seminary's main financial supporter, Lewis Tappan, asked about the controversy, Weld wrote a long letter.

"We believe faith without works is dead. We have formed a large and efficient organization for elevating the colored people of Cincinnati."[331]

Weld did keep a secret from Beecher. The students had made Lane Seminary a station on the Underground Railroad.

As Huntington Lyman later described, he had volunteered his prize fast horse for these covert ops. *"It was known that the horse might be taken without question by any brother who had business of Egypt."*

(The code words on the Underground Railroad often reflected the story of Moses leading the Israelites out of Egypt.)

"My horse was used hard. Between the Ohio River and the south end of the Underground Railroad we advised to send people over, not only with great speed, but also not even to tell each other when we did it."[332]

Weld kept leading the off campus activities. Beecher protested to the students, *"You are right in your views but unrealistic in your methods."*[333]

No one listened to Beecher. So he took a break from the seminary. Accepting invitations to preach across the country Beecher left the school trustees to deal with the student problem. He advised, *"If we do not amplify the evil by too much alarm, impatience or regulation the evil will subside."*[334]

School adjourned for the summer and the students continued their service. All summer long they worked in the community. The news spread across the nation. Southerners had renounced their status and inheritance to donate their time serving the black community.

The nation was surprised. Colleges were not supposed to discuss such a sensitive subject. How could this have happened? Beecher was confronted by very angry authorities.

America's leading religious leaders ordered Beecher to go back and keep his students under control. He knew that was impossible so he stayed away from his own school for a while.

The Lane Seminary trustees were worried. They could see the storm coming.

Many of them were local business owners. One was the Mayor of Cincinnati. They had a lot to lose.

Cincinnati's economy depended on trading with the South. The city's two hundred forty cotton gins and twenty sugar mills would shut down without southern products. Too many jobs were on the line. The board was furious at how these seminary students seemed too self-righteous to care about the rest of the town.

It was unheard of for the students to not only spend much of their free time in the city with their black friends, but to invite those friends to hang out with them on campus.

The board discovered that the students, *"Had brought a colored woman into church and seated her beside one of the most prominent white ladies in the city, in order to break down what they called a wicked discrimination in society."*[335]

The only board member who dared to side with the students was Asa Mahan. He described the uproar. *"Everyone was talking about it. Great excitement followed as the newspapers spread the facts before the nation."*

"Cincinnati had never been so convulsed before. The most influential citizens openly talked of sending an organized mob to demolish the buildings and drive the faculty and students from the grounds."

The danger even came home with him. He described, *"Our little daughters were playing in front of our home when a group of other children saw them and yelled, 'Their father is an abolitionist. Stone them.'"*

"Rocks were thrown and they fled, one of them suffering a heavy fall on the pavement."[336]

Mahan also felt the pressure to remain neutral on the subject. *"Christian brethren avoided me as though I had leprosy."*[337]

In addition to his duties at the seminary, Mahan also pastored a local church. When he denounced slavery from the pulpit, he started receiving visits from people in the town who insisted, *"It's unwise for you to jeopardize your influence to defend one single principle."*

Mahan replied, *"I had not so learned Christ (Ephesians 4:20)."*[338]

Mahan was summoned by the trustees to an emergency board meeting.

There he heard, *"A meeting had been held in New England by the presidents and leading professors of America's top colleges and seminaries."*

"They came to a unanimous decision that new rules must be passed to suppress all anti-slavery action. They agreed that Lane Seminary would lead and the others would follow their example."[339]

The board discussed how most of Cincinnati *"was looking at the seminary as a nuisance more dreaded than cholera or plague."*

Knowing that the campus had gotten involved in the Underground Railroad they worried about the *"Spirit of insubordination, resistance to law, and civil commotion fostered by the students."*[340]

One trustee warned: *"It's impossible for people who haven't lived in a slave state to understand the intensity of public opinion on the subject."*

"Once it gets stirred up, it's much worse than tampering with a hurricane or lightning."[341]

Mahan had a brutal argument with the board.

They vowed to crush the rebellion.

He disagreed saying, *"Ministers who shape their walk by public opinion only receive the reward of man. I'd rather receive the smile of God."*[342]

Mahan was promptly outvoted.

The board decided to expel Weld.

Mahan decided to warn him.

LANE REBELS

While school was out for the summer, Mahan went to Weld's home and tipped off the students to what the trustees had planned.

New rules would be strictly enforced, including one which would interfere with their Underground Railroad work. Horses would be banned from campus property.

Huntingdon Lyman blurted out, *"I can't believe they are treating us like mischievous boys who need parenting."*[343]

Henry Stanton spoke up, *"Then we'll leave and spread the whole matter before the public until their ears tingle. We will never surrender an inch of our principles."*[344]

When the next semester began in fall of 1834, the forty-six students returned. They were ready for battle.

Seventeen new students had also enrolled, after hearing about the controversy and wanting to join it.

Beecher was still absent.

The trustees cracked down on the students with harsh new rules. Free speech was denied. All off campus activities were forbidden.

The trustees accused the students of:

- *"Injuring the prosperity of the seminary;"*
- *"Acting without permission of the trustees;"*
- *"Causing a state of anarchy in the seminary;"*
- *"Being at war with the faculty;"*
- *"Socializing with colored people and"*
- *"Stirring up animosity, evil passions and bitter partisan prejudice."*[345]

The students were ordered to submit to the harsh new rules.
The students prayed about it and decided to revolt.

Together, fifty-one Lane Rebels signed a public statement.

"We withdraw from Lane Seminary because the school demands that we prostrate our principles to public opinion."[346]

Sacrificing their comfort, careers and social status, the rebels published a thirty-two page manifesto of why they revolted. *"We formed our group to operate against the system of slavery."*

"Is this a time to stop, when the pulpit is controlled, the press submits to power and conscience surrenders to convenience?"

"When the heart of the slave is breaking with the anguish of hope deferred and our free colored brethren are persecuted?"

"When the shepherds of God's flock, instead of carrying His lambs in their arms, tear them from the fold and hurl them to be torn by wolves!"

"Yet the church caresses these blood stained ministers! No! With all our hearts, minds and strength we answer NO! God forbid that we should abandon the cause of human rights!"[347]

The nation was shocked. Most of the Lane Seminary student body had walked out.

Suddenly the funding of the school was gone. Tuition was not being paid. Potential students, who had planned to enroll, suddenly changed their minds.

Beecher was furious, even though he was still getting his full salary. He had been embarrassed in front of the nation.

Beecher issued a statement blaming Weld for all the problems.

"We are disappointed with his insubordination and failure to follow our wiser guidance."

"Weld was a genius but uneducated. He would have made an incredible minister if his education had been better. He took the lead of the whole institution. The young men thought he was a god."[348]

Weld left Lane Seminary. Then he opened his mail to find a $1,000 check from a friend.

Lewis Tappan had sent a letter, asking Weld to find a new home for the rebels. Where should they go?

Many of the Lane Rebels wanted to stay and continue teaching in the community schools. Weld forwarded money from Tappan to pay their expenses so they could continue volunteering their time.

Some of them wanted to finish their education at a different school. Yet while many colleges would not even consider accepting these rebellious students, they soon received an unexpected invitation.

A new college named Oberlin wanted them. It had opened its doors about the same time as Lane Seminary, but struggled for students. So Oberlin's trustees visited the Lane Rebels and invited them to come finish their studies.

Twenty-eight students accepted the invitation, but they insisted on two conditions. Oberlin would have to appoint Asa Mahan as its President and start admitting African American students.

Oberlin agreed, even offering Weld himself a nice salary to come and teach.

Weld declined because he felt that God wanted him to do a different type of work.

He wrote back, *"God has made it clear to me that the priority of my life is abolition and the elevation of colored people."*[349]

Weld recommended his friend Charles Finney for the job. They hired him. And Oberlin became the first American college to admit both black and female students.

Future alumni would include famous people like Katherine Wright, sister of the Wright Brothers, who also generously supported Oberlin College.

Just as the Lane Rebels arrived to start classes at Oberlin, someone else showed up.

Beecher visited the Lane Rebels and begged them to return. He claimed the trustees had only acted to prevent violence from ripping the community apart. He promised to fire any instructors they didn't like.

They didn't listen. Pierce wrote to Weld describing what happened. *"Lane's faculty have done what they could to reclaim us, but all in vain. We are reckless."*[350]

James Thome was another Lane Rebel who transferred to Oberlin. He wrote, *"The Lord has led me by strange paths ever since I entered Lane Seminary."*

"My feelings and friends changed. Thank God my calm life has been thrust into conflict. Though I've lost people's approval, I've found friendship with Heaven and peace of mind."[351]

Thome's slaveholding family thought he was crazy. They visited and tried to talk him out of his strange new beliefs.

He stood his ground. He made them think about what they were doing. Slowly their minds began to change.

Within two years, Thome's father had freed everyone on his plantation. One of the happiest days in Thome's life was when those former slaves came and thanked him for doing the impossible.

James Bradley, the African American Lane Rebel, had also transferred to Oberlin. The local newspaper, *Oasis*, asked him to write about his experience.

He did, describing the excitement of how things had changed. *"My heart overflows when I hear what is being done for the poor broken hearted slave and the free men of color."*[352]

After that article was published, Bradley disappeared from the historical record.

Meanwhile, Weld was back on the road, traveling and preaching as a full-time employee of the American Anti-Slavery Society.

Tappan paid Weld's expenses so he could do what he did best. Preaching all over the west. When people wondered why Weld would preach in the west instead of the south, Weld explained that this was how to turn the tide of the nation. Convert one small town at a time and soon there would be an unstoppable force of abolitionists across the nation. It was just a matter of time and labor.

He would spend several days or weeks in each town, preaching night after night, until public opinion was on his side. Then he would organize an abolition group to carry on the work after he left. These little abolition groups would make a big difference. Many of them petitioned Congress. Others rallied voters to elect good candidates into office.

As the Bible tells the story of John the Baptist preparing the hearts of the people before Christ came, Weld was preparing the hearts of the nation. The day was coming when their votes would determine the outcome of the Civil War.

Weld had a powerful gift. Few people could move an audience like he did.

Birney described, *"No one could speak like Weld. He touched the heart, moving his listeners through a wide range of feeling. They wept or laughed with him. He had no selfishness, hatred or sharpness. Only compassion."*[353]

A newspaper reporter, who listened to Weld for two weeks straight, still struggled for the right words to describe it because *"You can't print thunder."*[354]

One of Weld's southern converts was seventeen year old James Davis. He wrote to Weld, *"Before I became an abolitionist, I had money as I wanted. My father was a man of influence and I was respected and loved too."*

"Now the people of Kentucky shun me as they would a rattlesnake. How do I feel? Independent! My trust is in God."

Leaving home to move to a free state, Davis told Weld, *"If I never see you again, just know that I love you like my heart's blood."*[355]

Future Oberlin College President Fairchild wrote of Weld, *"I have seen crowds of bearded men held spell bound by his power for twenty straight evenings."*[356]

Soon Weld had another reputation. The more effective he was, the more the merchants tried to stop him. They didn't want him interfering with their sales to the South of everything from silk to tools. They had been threatened by the slave owners to shut down abolition meetings or lose their customers.

Violence started to happen, wherever he spoke. Quickly he became, *"The most mobbed man in America."*[357]

The Richmond, Virginia newspaper, *The Whig*, didn't even try to hide their anger. They printed a furious editorial. *"If the northern people want to keep their lucrative trade with the South they must start hanging these fanatical abolitionist wretches."*[358]

So merchants hired thugs to disrupt Weld's meetings. Chaos began to greet Weld wherever he went. People were sent to start riots in town after town. People were injured. Property was destroyed. Law enforcement turned a blind eye.

Weld described one of his events. *"I had hardly begun to speak when a mob gathered with tin horns, sleigh bells, drums, etc and bedlam broke loose."*

Rocks were thrown through the windows. One rock struck Weld so hard he had to pause *"Until the dizziness ceased."*[359]

Weld kept preaching. The audience helped him by taking off their coats and covering the windows.

Outside the meeting, the audience's horses were assaulted.

Weld described, *"Snuff was thrown into women's eyes. Innocent sitting hens were disturbed in their nests and the eggs thrown into the meeting. Cannons were fired."*[360]

Yet, *"The Lord restrained them. Not a hair of my head was injured."*[361] The men in the audience also protected Weld, making sure he traveled safely back and forth to where he was staying, the entire time he was in the town.

Weld had a powerful weapon against the mobs. Patience. When they disrupted his meetings, he would stand there patiently until they got tired and went home.

Then he would hold another meeting the next night. And the next and the next. He kept speaking every night in the town until he had worn out the mob. This courage impressed the audience. The more rioters disrupted him, the more people rose up against oppression.

When Weld asked them to get involved, they did. Town after town organized abolition groups.

Meanwhile, the proslavery side mobilized its efforts to counter the revolution sweeping the nation. The *Anti-Fanatical Association* was even formed in Louisiana to fight back.

On September 17, 1835, rewards of $500 were offered for the arrest and prosecution of any abolitionist leaders on charges of *"Attempting to excite violence."*[364]

The violence spread to New York City. One day as Lewis Tappan was going home, he saw a newspaper headline. His home had been attacked by an angry mob.

He hurried home and found his house heavily damaged by the mob. He wrote to Weld describing, *"There was the most wanton destruction of property you can imagine. Pray for us. If we die here at our posts don't desert the anti-slavery cause which is dear to the heart of our Savior."*[365]

While Lewis Tappan paid a heavy price for being an abolition-ist, God rewarded his faithfulness. Even though his business suffered during an economic downturn, that would give him an idea for a new business. He created the credit rating agency, which became Dun & Bradstreet. (That company still exists today and is deeply respected.)

As more and more people asked Beecher what had happened to his former students he replied, *"They think they do God service by butting everything in the line of their march which does not fall in or get out of the way. I hope they will calm down and stop criticizing."*[362]

Mainstream churches began issuing warnings to their people to avoid any connection with these crazy abolitionists.

The General Conference of the Methodist Episcopal Church—held in Cincinnati in 1836—warned their people about the *"great excitement"* shaking the country. *"As your pastors we urge you to avoid supporting the abolitionists and completely refrain from this disturbing subject."*[363]

As the violence escalated, even Charles Finney worried about Weld. Maybe he had gone off the deep end. Yes, Finney agreed that slavery must be destroyed. But they were going about it all wrong.

Couldn't Weld see the coming storm? Finney wrote a letter telling Weld to spend more time in prayer and fasting. Wait for God to change hearts. Finney even accused Weld of pride by saying, *"Be careful or you will be spoiled by the idea of your own importance."*[366]

Demanding that Weld change his methods, Finney told him to be silent and preach only revival meetings. *"Aren't you afraid that we will head straight into civil war unless your methods of abolitionizing the country are changed?"*[367]

Insisting his ministry methods were the only way, Finney pushed the Lane Rebels to change.

At Oberlin, Finney used his classroom time to teach the students to stop talking about freedom and only preach salvation and relationship with God. Nothing else.

If they could save enough sinners, then the slaveholders would change their hearts on their own. Finney complained, *"Very few of the abolitionists are wise."*

He worried about how these *"Reckless, critical, uncooperative men"* were distracting the country from the possibility of nationwide revival.[368]

Several Lane Rebels wrote to Weld and asked for his opinion. *"If we ever needed your counsel, we need it now. Should we preach abolition or religious revivals? Can't we save souls and save the slaves too?"*[369]

Weld quickly wrote back, urging them to seek God for the answer. Then he dropped everything to personally visit Oberlin.

Having a long, private talk with Finney, Weld pointed out that God had called different people to preach different things. He respected the successful revival ministry Finney had once had. But God was moving in many other ways.

Weld said Christians needed to focus on the area of their own calling while still supporting the areas of other people's callings.[370]

Then Weld confronted Finney for not doing enough for both abolition and civil rights. Weld made his opinions very clear.

"The reason the abolition movement isn't gaining more ground is because too many of our prominent men are too afraid of public opinion. Yet how can we cringe at the obstacles in our way when God has decreed the ultimate triumph of this cause?"[371]

Finney listened and softened a little. In his public services he began to speak out more against slavery. Yet he refused to change at the seminary.

Convinced that his methods were the only way, Finney forced a church split. He divided the Lane Rebels, making them choose between training under him or Weld. And Finney got very angry at the rebels like Henry Stanton who disagreed with him.

Meanwhile, the Lane Rebels got together and discussed the situation. They prayed about it, met with Weld about it and then decided to continue fighting for freedom in their ministries.

Following their hearts, they felt the Lord leading them to continue their work. It was perfect timing for a big opportunity to come their way.

Weld had studied how Jesus had sent out the disciples to preach the Gospel. At one point, Jesus sent seventy people to reach more towns than one person could do alone.

Weld had an idea to train seventy people to travel the country and preach. Tappan's Anti-Slavery Association was willing to pay their salaries so they could travel to different parts of the country at the same time, all preaching abolitionism.

Weld was humble enough to realize he couldn't reach the nation by himself. The seventy men and women would go much further than he could alone.

Out of the twenty-eight rebels at Oberlin, eleven joined Weld's group. Weld also found other people to join.

Before sending them out, Weld took extra time to train them on everything from argument skills to dealing with hecklers.

He even taught them the easiest way to clean up, if they were tarred and feathered.

Most of his time, he taught them the Bible verses against slavery. They needed to be able to debate anyone, especially people more educated than themselves. Then Weld sent them out to spread wildfire across the country.

His team of seventy would turn America upside down.

Yet they would face heavy opposition. Criticism rained down on them from all sides. They had a hard time finding places to preach. Many churches would not allow them to use their building. Many pastors did not want to get involved. Shutting their pulpits to abolitionist preachers, they denounced these radical fanatics.

Still the team of seventy found other places to preach. However, thugs also found them. Well organized mobs were sent to disrupt abolitionist meetings. They weren't just hired thugs. Even local pastors attacked the abolitionists.

At one town he visited, James Thorme couldn't believe his eyes. Several people were threatening him, including the local Methodist pastor who threatened him with a bat. He was banned from speaking at any of the local churches. No one else wanted him to use their venue for his event.

When he finally found a place to hold the meeting and began to speak, *"We heard the yell of the mob outside. Twenty-six glass window panes were broken. Egg shells, whites and yolks flew on every side. No one escaped the splattering. I tried but couldn't clean off the stink."*

Thome wrote to Weld how the end result of the meeting was the town had formed a new abolition group. All the egging had only resulted in *"Hatching abolitionists fast."* He bragged, *"I was as brave as a warrior."*[372]

Henry Stanton was also getting used to the rising violence. *"When we started, slavery held the church and state by the throat. James Birney and I traveled east and encountered the first of my two hundred mobs."*

"When I went to Norwich and spoke in the town hall, stones were thrown until every window in the room was broken. We adjourned until planks were nailed across the windows to protect me from the missiles of the mob. Then I spoke for two hours. I was mobbed in every state from Maine to Indiana."[373]

Did that slow Henry Stanton down? Not a bit. He kept going and going. When he was asked to testify before the Massachusetts State Legislature, he described the violence they were experiencing.

Yet he promised they would never bend to pressure. *"Even if our nation was ripped apart into ten thousand pieces, antislavery agitation would hunt down the slaveholders and scatter upon them the living coals of truth from God's Word, 'Woe unto him that uses his neighbor's labor without paying wages (Jeremiah 22:13).'"*[374]

Another Lane Rebel, John Alvord, quickly discovered how dangerous it was to be part of Weld's seventy. As soon as he hit the road, the mobs didn't even wait for his meetings to start violently attacking him.

One night he woke up to hear thugs breaking into his hotel room. They dragged him from his room and forcibly threw him out of town.

Alvord was very grateful. They had only threatened to tar and feather him. He could have been hurt much worse.

He kept traveling and speaking across Ohio. Riots followed him everywhere but the local people protected him.[375]

By the end of the year, the Lane Rebels had turned the entire state of Ohio upside down. It would become so full of abolitionists that people automatically assumed anyone from Ohio was antislavery.

Lane Rebel Amos Dresser realized that when he made the mistake of traveling through Nashville, Tennessee on his way to visit family. His name had been published on the list of Lane Rebels. No sooner had he checked into his hotel room than the town people realized, *"He's from Cincinnati, he must be an abolitionist!"*

The town mayor, John Erwin, ordered Amos to appear before the judge. Amos' suitcase was searched. Abolitionist pamphlets discovered.

The judged examined Amos' private diary and determined, *"It is very hostile to slavery."*

When he was finally allowed a chance to defend himself, he dared to call slavery a sin. That made the audience furious.

Amos was sentenced to twenty lashes. As he was dragged from the courtroom, he noticed the crowd watching him was full of angry people. They were not satisfied until they heard, *"The sound of cowhide upon my naked body."*

Amos prayed for them, while they hurt him until the crowd roared, *"God damn him, stop his praying!"*[376]

Then the crowd forcefully expelled him from their town. Amos went back on the road, preaching the gospel again.

Fellow Lane Rebel Marius Robinson was also literally thrown out of a town. Dragged out of bed in the middle of the night, he was stripped, tarred, feathered and dumped by the side of the road.

He was found by a farmer, who cleaned his wounds. Dazed but determined, the very next day he gave another abolitionist speech at another town. Writing to Weld, he reported that being, *"Considerably bruised and sore"* helped him *"lecture better."*[377]

Another Lane Rebel willing to face the danger was William Allen. After signing the rebel manifesto, he was warned not to go home.

His slaveholding family and Alabama hometown were furious. Allen bragged to Weld, *"They've said they'd cut my throat."*[378]

Those neighbors had plenty of motivation to do it. Allen had become one of Weld's seventy, risking his life to preach around the country.

He pressed forward even when his own father criticized his work as *"A miserable project characterized by reckless ambition and rank pride,"* and demanded that he *"Return to his duty at the seminary."*[379]

Instead of listening to his family, Allen preached everywhere from Ohio to New York. Then he felt the Holy Spirit guiding him back towards Illinois.

Allen would transform the state of Illinois. He wrote to Weld, *"The people are mostly anxious to hear and ready to be converted."[380]*

In Illinois, he met a powerful pastor, Owen Lovejoy, who preached the Bible on Sundays and ran the Underground Railroad on Mondays. Lovejoy was changed by Allen in a completely different way.

Allen convinced Lovejoy to run for Congress. How it happened would be described years later by Lovejoy in a letter to Weld.

"I had heard of William Allen as an abolitionist whom mobs couldn't scare, but hadn't seen him until one day a tall man touched my shoulder with a giant's grip as he said, 'Mr. Lovejoy you ought to be in Congress. You're just the man to lock horns with those bellowing slaveholding animals.'"

"I replied, 'Nothing short of a miracle could compel my congressional district to send an abolitionist to Washington.'"

"Allen said, 'I'm going to stump your district. I'll lecture wherever I can get a church, a hall, a schoolhouse, a private home or a barn. If I can't get them, I'll preach abolition under God's open air until the thing is done.'"

Lovejoy marveled at how *"Allen did just that. He actually abolitionized my congressional district and sent me to Washington."[381]*

Illinois would send another very important person to Congress. Before the Lane Seminary debate, Lincoln had run for office in 1832 and lost. Then in 1834, while the Lane Rebels were publishing their manifesto, Lincoln ran again and won for the first time.

Lincoln was elected to the Illinois State Legislature in 1834 and re-elected in 1836. Then Lincoln made his first antislavery speech in the State Legislature session on January 11, 1837. Then Illinois sent Lincoln to Congress in 1846.

By the time of the famous Lincoln/Douglas debates in 1858, the Lane Rebels had affected so much of America that at one point during the debate, Stephen Douglas accused Lincoln of trying to imitate the abolition preachers.

Owen Lovejoy was with Lincoln during the debates. By that time they had become close friends.

At one point when Douglas insulted Lincoln and Lincoln jumped up to say something he would have regretted, Lovejoy pulled Lincoln back down and saved him from embarrassing himself.

Lovejoy would also convince Lincoln to join the newly formed Republican party. Together they would work in Illinois politics to put antislavery candidates in Congress.

Their work was considered offensive to some. Abolitionists were still mocked, scorned and accused of extremism.

Levi Coffin was running the Underground Railroad in Indiana at the time. He remembered, *"Abolitionism was very unpopular."*

"Antislavery lecturers began to travel my state and stay at my home. But as the movement gained strength, the opposition to it became more powerful."

"Becoming an abolitionist tried a man's soul as bats, stones and rotten eggs were the arguments we had to face. My wife's nephew had two of his front teeth knocked out by a brick thrown by the mob. Yet public opinion in my neighborhood became so strongly antislavery that I often kept fugitives openly at my house without any fear."[382]

Weld's group of seventy would have a profound effect on the nation. Their influence lasted many years after their travels ended. Just one example was what happened to Jonathan Blanchard.

While traveling through Pennsylvania, he preached at a small town. The people of that town told him about a grouchy old lawyer who lived there. The people told him not to even bother trying to convert that lawyer to abolitionism. The lawyer was a stubborn man who had defended slaveholders.

Jonathan didn't listen to them. He developed a lifelong friendship with that lawyer. One thing led to another and that lawyer would be elected to Congress. He would lead the battle for civil rights. Today that lawyer is remembered in history as Congressman Thaddeus Stevens, the driving force behind the Thirteenth and Fourteenth Constitutional Amendments.

WELD BECOMES A WRITER

While Weld's team was traveling and preaching across the country, Weld was also traveling and preaching. He worked so hard that he lost his voice. When he could no longer speak, he focused on writing. He moved back to the New York Anti-Slavery office and became head of the publishing department.

Day after day, he wrote pamphlets on the major issues that everyone was discussing. These pamphlets would go across the country, reaching even more people than Weld did himself.

One of his most powerful writings was *The Bible Against Slavery*. In this book, Weld went verse by verse to demolish proslavery doctrine that had infiltrated the church. He had several interesting points.

Weld wrote how God gave the death penalty to slave traders in Exodus 21:16 (KJV). *"Whoever steals a man and sells him, shall surely be put to death."*

Then he wrote about how the Ten Commandments included:
"Thou Shalt Not Steal."

So if God forbade stealing one item, how could God condone stealing someone's whole life under the system of slavery? Weld went through verse after verse, showing the Bible taught people should be paid for their labor, not forced under the evil system.

He wrote, *"Slavery is the highest violation of the Eighth Commandment. To take from a man his earnings is theft. But to take the earner is a life-long theft."*

"The Eighth Command forbids stealing and the Tenth Command forbids coveting, protecting every man's right to himself and his property."[383]

Weld also answered the question that has often been raised in modern times.

Today's atheists have used Exodus 21:20-21 to accuse the Bible of allowing masters to beat their servants as much as they wanted, as long as they didn't kill them. That is totally wrong.

Weld showed the context of Exodus 21 was describing disability law. In the *"Case of personal injury, the injurer is to pay a sum of money."* The passage talked about how if a servant lost an eye, he was to be compensated for the injury.

Then just a few verses earlier in Exodus 21:12, the Bible gave the death penalty to anyone who strikes and kills a man.

In that same chapter, the Bible also gave the death penalty to slave traders (Exodus 21:16). Plus, this came right after Exodus 20, when God gave the Ten Commandments.

Weld wrote that the same God who *"Thundered from Mount Sinai, 'Thou Shalt Not Kill,'"* would not turn around and approve of murder by a master. This passage of the Bible was actually giving the slave's life the same value as the master's life.

Exodus 21:20-21 described how the punishment for murder applied to slaveowners too. That the Hebrew word *"nakam"* was typically translated as *"vengeance"* or *"revenge"* in other parts of the Old Testament.

The true meaning of this passage was: if the master beats the slave to death, then slave's death had to be avenged, because the slave's life was just as valuable as the master's life.

Weld wrote Exodus 21:20-21 was actually saying, *"If a man smites his servant and the servant dies, the death shall be avenged."*

"The death of the servant shall be avenged by the death of the master."[384]

If the slave survived the beating and lived, the master would still have to pay him compensation for his injuries. The punishment would be a monetary fine instead of the death penalty. This passage was also describing the difference between manslaughter and first degree murder. There were different punishments for each, depending on the severity of the crime. The Bible was describing how the master should be punished according to how severely he had hurt the servant.

Weld also wrote another powerful pamphlet with detailed research on how Congress had power over the District of Columbia and could destroy slavery in D.C., even if there were too many obstacles for it to abolish slavery elsewhere. This was a major disputed point of the time since D.C. was the only place Congress could pass an abolition law without interfering with state's rights.

Weld's voice would be heard. Later during the Civil War, Congress would pass a bill to abolish slavery in D.C.

Weld's writings had such a powerful effect, soon he realized, *"Ever since I stopped speaking, I've been able to do twice as much for the cause."*[385]

Weld's writings touched so many people that when the British novelist Charles Dickens visited America, the one person he wanted to meet was Weld.

The tide was turning. Town by town, America was changing.

Eyes were opened and people rebelled against the propaganda that had once been shoved down their throat. People rose up, willing to sacrifice for truth and justice.

They formed groups all over the west. These groups didn't just pray. They acted. They mailed abolition pamphlets to southern plantations.

They organized new Underground Railroad stations. They voted into Congress new antislavery politicians. They flooded Congress with half a million signatures on the petitions that John Quincy Adams used to fight the system.

This was the chance that Adams had been looking for. Ever since being elected to Congress in 1830, Adams had planned how he would attack slavery.

When the Lane Rebels published their manifesto on December 15, 1834, it caused the nation to begin discussing the forbidden topic.

Months later, as Weld's group of seventy were changing the nation, it was perfect timing for Adams to begin his one man war in Congress on January 4, 1836.

The proslavery power fought back.

On February 1, 1836, Rep. Hammond gave a speech in Congress warning the South of the rising movement. *"There is no doubt of the deep, penetrating, uncontrollable excitement which is shaking the free states."*

Hammond showed Congress copies of new abolitionist publications. He was highly disturbed that 350 new abolition groups had sprung up since 1832. *"All their meetings and publications are meant to teach the slave to cut his master's throat."*

Hammond warned that emancipation would only lead to a violent race war between blacks and whites and *"This republic will sink in blood."*

He insisted, *"We love this Union, but we will never sacrifice our rights on its altar. This question can only be settled by blood."*[386]

By 1840, the wildfire had spread so far that Rep. Charles Ingersoll of Pennsylvania stood up in Congress to complain now America had two thousand new antislavery groups. He warned the South to join together to *"guard its rights"* against this dangerous trend.[387]

Senator (and future Confederate President) Jefferson Davis made a furious speech on the Senate floor against, *"The intrusion of strangers into this most delicate subject. They have raised a storm they cannot control. Your loyalty to the Constitution should restrain you from trampling on our rights. Keep down this excitement!"*[388]

The wife of Jefferson Davis, Varina Davis, complained: *"The growth of the antislavery movement in the free states had been rapid and alarming."*

"Its leaders were becoming more and more aggressive by helping fugitive slaves escape, by the circulation of the most offensive petitions and by public speeches denouncing the South. Northern people were being prepared to insist on domination instead of equality of constitutional rights."

"Day after day in Congress, southern men were forced to suspend the business of the country to hear themselves insulted by petitions. The South began to realize the only remedy left was withdrawal as the irreconcilable conflict began."[389]

One after another, the legislatures of southern states passed resolutions demanding that northern states stop this nonsense.

North Carolina demanded: *"New laws prohibiting any publication which might make our slaves discontent."*

Alabama wanted: *"New penal laws to put an end to the evil work."*

Georgia demanded: *"That the North crush these traitors."*[390]

Congress joined the fight and pushed back on these abolitionists.

Senator Robert Goldsborough of Maryland said the abolitionists were only making things worse for the slaves. He demanded Congress pass new laws crushing this disturbance *"That they might proceed without excitement."*[391]

Senator William King of Alabama demanded, *"Deluded followers of leading abolitionists be rebuked and discouraged."*

He was shocked at how these groups *"Had increased beyond all calculation."*

He warned, *"Their request is unreasonable and cannot be granted as it would violate the Constitution. It is impossible for abolitionists to benefit the slaves, even if they could succeed in changing Congress."*

King closed his speech by *"Relying on his northern friends to use legislation and newspapers to push back public opinion. Put down the dangerous spirit of abolition!"*[392]

Congressman Lawrence Keitt of South Carolina gave a long speech also complaining of how *"Instead of being crushed out, organized abolition groups are getting more focused and aggravating. Abolitionism now controls the authorities of ten states."*

He wondered why the South should submit to *"This vulgar tyranny"* when they had more than enough resources to resist any attempt to interfere with slavery. Reminding everyone that the South was paying considerable taxes on its annual exports of over $128,000,000 of cotton, he stated, *"The South furnishes more than two-thirds the exports of the country."*

"Aren't the resources of the South abundant to establish an independent government? The loss of the cotton crop would cover England with blood and anarchy and shake down the strongest thrones in Europe. The South has nothing to fear."[393]

Congressman Wilson Lumpkin agreed. He told Congress that if the northern states couldn't keep the abolitionists under control, then it would be the duty of the southern states to *"Punish the abolitionists in the most ideal ways."*[394]

Abolitionists responded by getting even more obnoxious. They mailed Weld's abolition pamphlets to slave owners.

Slave owners were so enraged by this junk mail that they broke into the post offices, ripped open mail bags and burned any abolitionist literature they found.

It kept coming. Northern people kept mailing tracts to southern plantations until President Andrew Jackson ordered Congress to pass *"severe penalties"* against abolitionist literature.

Southern postmasters publicly announced they would not deliver this mail as the South was too tender to receive such offensive opinions.

Abolition tracts kept pouring into the mailboxes of southern plantations. Once again, President Andrew Jackson ordered Congress to pass laws to stop it. He complained about the *"Painful excitement in the South caused by misguided people mailing provocative literature intended to bring the horrors of a slave revolt."*[395]

Senator Calhoun presented a bill, *"Making it unlawful for any postmaster to deliver any pamphlet, newspaper or other paper touching the subject of slavery."*

He claimed the bill was necessary to prevent slavery from being abolished which would *"Engulf the nation in a sea of blood."*[396]

When the bill failed to pass and Congress tiptoed around the subject, southerners again broke into their local post offices, ripped open mail bags and burned every letter from the North that they found.

Postmasters responded by refusing to deliver any antislavery mail.

Some still got through. One pamphlet reached a Baptist pastor living in South Carolina named William Brisbane.

Born into a family with a long history of owning plantations, Brisbane saw nothing wrong with being both a Baptist pastor and plantation owner.

Until he read Weld's pamphlet. One day he opened his mailbox to find that someone had anonymously mailed Weld's pamphlet to him. Deeply offended by it, he quickly threw it in the trash. But then he couldn't quit thinking about it. Didn't he know the Bible better than Weld?

After all, Brisbane edited a Christian newspaper. Already he had written article after article defending slavery. Surely he could disprove one crazy pamphlet.

Yet when he opened Weld's pamphlet, it opened his eyes. It made him take a closer look at the Bible. Brisbane began examining specific Bible verses in the original Hebrew and Greek. That's how he realized the Apostle Paul had never told slaves to obey their masters.

The correct translation was telling employees to submit to bosses and bosses to pay fair wages. The New Testament actually forbade employers from denying wages to their staff, which alone attacked the whole system of unpaid slave labor.

The more Brisbane thought about it, the more he realized the truth. *"Discovering my error, I found myself perplexed with my own arguments."*[397]

He realized, if he really wanted to live his life according to the Bible, he was going to have to turn his entire life upside down.

Brisbane did. First he freed his slaves and paid for them to move to a free state. Then he wrote a powerful book about his conversion called *Slaveholding Examined in the Light of the Holy Bible.* In the book, he described how *"I know the struggles in a slaveholder's mind and how they mistake the Bible."*[398]

That cost him his family, finances and friends. His life was threatened. He was complexly ostracized from the community, leaving him with no choice but to move to a free state.

He moved to Cincinnati, Ohio and became friends with Levi Coffin who was working on the Underground Railroad.

Together they ran the Underground Railroad in that area until the Civil War. After the war, Brisbane was offered a new job.

The federal government sent him back to his old neighborhood to collect unpaid taxes. He would end up selling the plantations of slave-owners who had refused to pay property taxes during the war.

Afterward he became a Baptist pastor in Wisconsin. While he paid a heavy price to do what he knew was right, he would never regret his decision.

So what happened to the Lane Rebels?

All of them remained faithful to their principles.

Most of them ended up in full time ministry, pastoring churches. Finding their own ways to preach the gospel, they turned the hearts of their congregations towards freedom. Their communities changed. Public opinion changed. More people voted for anti-slavery politicians.

Lane Rebel Uriah Chamberlain became a pastor by day and conductor on the UGRR at night. Yet his hard work was punished. He was caught and fined $50,000. He refused to back down. He boldly proclaimed, *"God has anointed me to preach deliverance to the captives even if it costs me."*[399] He appealed the fine until it was overruled.

William Allen left the ministry to become a justice of the peace. It was the perfect cover for his real occupation. He operated a very successful Underground Railroad station in Geneseo, Illinois.

James Thome worked with several other Lane Rebels to organize the American Missionary Association.

He told Weld, *"These last three years I've enjoyed walking into a new world. Trouble feels only like a gently falling snowflake. No longer can I do anything else. I want to know nothing but Christ, as I promote His cause."*[400]

John Alvord was given a very special job after the Emancipation Proclamation. When the federal government formed the Freedmen's Bureau to assist the newly freed families, Alvord was appointed the superintendent of education. Fellow Lane Rebel George Whipple assisted him in organizing new free public schools in the South.

David Ingraham became a missionary to Jamaica. For four years he served in full time ministry, launching new schools there.

The hard work broke his health. In 1841, he came back to America to visit Weld. The day after arriving at Weld's home, Ingraham collapsed and died of exhaustion. His ministry in Jamaica lived on.

Henry Stanton got involved in politics. His efforts to elect new anti-slavery Congressmen would birth the Republican Party.

Yet he would remembered by history only for being the husband of women's rights activist Elizabeth Cady Stanton.

In 1840, when he took her to the World Anti-Slavery Convention in London, women were not allowed to be seated. That discrimination would inspire Elizabeth Cady Stanton to go home and launch the women's suffrage movement. She led it for many years, inspiring many other women, including Susan B. Anthony.

WELD WORKS FOR ADAMS

Meanwhile, Weld also served his country by getting involved in politics. After Weld had lost his voice and was no longer able to preach on the road, he received a very unexpected invitation.

Congressman Joshua Giddings wrote to him, asking for help. John Quincy Adams was fighting the *"gag rule"* in Congress. Giddings was helping him, but the battle had become so intense, more help was needed.

Could Weld come to Washington D.C. and help them research the law? They needed heavy legal documentation to prove the *"gag rule"* was unconstitutional.

Weld moved to Washington D.C., staying at the same boarding house where Giddings lived. This boarding house would later become known as the house of abolitionists because it was the place where the abolitionist Congressmen met to plan their strategies.

The lady who owned this boarding house, Mrs. Spriggs, also owned slaves who worked there. That helped Congressman Giddings hide how he was running a very active station on the Underground Railroad out of the boarding house.

Weld helped him with it. Together they worked with a local Pastor, Charles T. Torrey, who focused on rescuing the slaves owned by U.S. Congressman.

They were able to help many slaves escape. So many of the slaves, who worked at the boarding house, disappeared that she sent the rest to live elsewhere. Then she hired new people to work in their place. Weld wrote home to family, how proud he was their work had transformed the boarding house.

The success came at a heavy price. For running the Underground Railroad, Charles T. Torrey was caught, arrested and sentenced to six years in prison. He would die in prison several years before the Civil War. His sacrifice would inspire many other abolitionists, including John Brown.

Meanwhile, the work continued. John Quincy Adams was thrilled to have Weld working for him. When they met in person, Adams told him how much he had enjoyed reading Weld's books.

"I know you well by your writings."[401]

Weld became the first Congressional staffer in history. Adams even reserved a desk for him at the Library of Congress. Weld spent many hours there, researching American history and the laws to prove slavery was unconstitutional.

His research was utilized by Adams during debates in Congress. When Adams was put on trial in Congress for treason, Weld worked on the legal research.

When Adams won that battle, Weld wrote,

"The triumph of Mr. Adams is complete. This is the first victory over the slaveholders in a (government group) body since the foundation of (America's) government.....The slaveholders had been so horribly burned by the hot iron that they took special care not to make any more errors."[402]

Weld loved when he was able to sit in the visitors gallery of Congress and watch Adams, *"Dodging points of order, creeping through this rule and skipping over that, all the while preaching tiny speeches which were so short and fast that though they were out of order, nobody could call him to order and when they did he would say 'My speech is done.'"*

"Then Adams sat down, half rabid and half laughing. He had had his way even with the opposition of two hundred to one."[403]

Weld was impressed. *"The energy with which he speaks is astonishing. Though seventy-five years old, his voice is one of the clearest and loudest in the House."*

"After he spoke in the House for three hours with great energy against his assailants, I was afraid he had tired himself out. In vain I tried to stop him, telling him he would need all his strength for the next day. He still went on speaking for an hour."[404]

Helping Adams gave Weld a brilliant idea for how to fight the battle. Day after day Adams had been wrestling in Congress against the powerful proslavery propaganda machine.

The propaganda was so strong that even today some people still believe the same propaganda which was being spread years before the Civil War. For example, most of the country had been led to believe the slaves were well treated. That the masters were kind people just trying to help their workers. Nothing was further from the truth. Yet how could Weld prove it to the world? How could he show how bad slavery really was so public opinion would turn against it?

When Weld's work in Washington ended and he moved back home to the New Jersey area, he began working on a new book. First he went down to the local library and asked if they had any southern newspapers.

They did. Weld offered to buy any old copies of these newspapers when it came time for the library to discard them to make room for the new ones. Over the next several months, Weld purchased twenty thousand copies of southern newspapers.

He closely studied these newspapers, writing down detailed notes on everything from ads for runaway slaves, to court cases and relevant news stories.

His idea was that since these newspapers had good reputations for reporting the facts, then it was the facts themselves that could be used to destroy the propaganda.

If he could just compile enough evidence from reputable sources, then he could prove how evil the system of slavery really was by letting these newspapers speak for themselves.

As Weld began putting his book together, he also wrote to the Lane Rebels who had grown up in slaveholding families. He asked them to write about their personal experiences. How their hearts had changed enough to repent of slavery. Many of them wrote back with powerful testimonies which were included in the book.

Then Weld researched what life was like for inmates in local prisons. His idea was to do a side by side analysis, proving how even as bad as prison conditions were at that time, life in prison was actually much better than on the plantations.

Finally, Weld did extensive historical research to show that throughout history, the tyrants of the world had always claimed to be good masters, even while they were abusing their people. By comparing the modern propaganda with the propaganda of ancient times, he showed how evil it really was.

When Weld finished the book, he had no idea he had just written a major bestseller. *American Slavery As It Is: The Testimony of a Thousand Witnesses* sold over 100,000 copies.

That's a remarkable achievement considering even today, over 150 years later, books only have to sell 5,000 copies to be considered a best-seller!

Weld's book would have a powerful influence on the nation. It would turn many hearts against the evil system. And it would reach someone with an interesting connection to history.

Harriett Beecher Stowe was the daughter of Lyman Beecher. Long ago when the famous debate had happened at Lane Seminary, she had attended and became a hardcore abolitionist.

When she heard about Weld's book, she bought a copy. Reading it inspired her to write something.

She would sit down and write one of the highest selling novels in American history, *Uncle Tom's Cabin.*

Her book had such a powerful effect on the nation, that during the Civil War, when she was invited to the White House, Lincoln asked if she was the little lady who had started the big war.[405]

Harriett also received a lot of criticism over her book. Lots of people were offended by it. The black community felt she didn't understand their right to stay in America because the book ends with the main character traveling back to Africa.

Meanwhile, the proslavery power accused her of making their system look bad, when they thought they were good to their slaves.

Harriett fired back by writing another book, *Key to Uncle Tom's Cabin.* Her second book pointed to Weld's book that had inspired her and the eyewitness testimony in Weld's book, proving how evil the system really was.

Weld continued fighting the battle until it exploded into the Civil War. By that time public opinion had turned so strongly against slavery, most of America was eager to destroy it. From those small towns—that Weld had believed would hold the future of America—would come people who would make a big difference in their own way.

There would be countless stories of bravery during the war. One was the Union Army regiment from Racine, Wisconsin.

They were nicknamed the abolition regiment because they developed a reputation for protecting fugitive slaves who came to them for help.

During the war, their leader Colonel Utley wrote, *"Slave catchers follow us day and night, determined to break down the principle upon which we stand."*

According to Col. Utley, one day, *"A boy came into our lines, cold, barefoot, ragged and hungry, asking food and shelter after having spent days in the woods living on black walnuts."*

Soon after, the master showed up with orders from General Gillmore to surrender the boy.

Col. Utley replied, *"I will see you in hell first!"*[406]

The master sued in court and won a $1,500 judgment against the Colonel. Then Congress passed legislation that the military did not have to turn over fugitives.

Colonel Utley's fine was reduced to $700 in court costs. He proudly paid it.

That same regiment was also proud to rescue an eighteen year old girl. She ran away from a Kentucky plantation, reached the camp and begged for help. Shortly afterward, her master showed up to demand they turn over his *"property."*

The soldiers refused. She was hidden in a wagon and driven by the soldiers over one hundred miles to a station on the Underground Railroad run by Levi Coffin.

He wrote, *"These brave young men telegraphed to Racine, Wisconsin and made arrangements for their friends there to receive her. They told me this was one of the happiest moments of their lives."*

The soldiers went back to their camp and reported, *"The Lord was truly with us on that journey."*[407]

When the Emancipation Proclamation was issued, it seemed almost like a dream to Weld.

This was the day he had fought for. This was the day that had seemed impossible for so long.

Once the victory was won, Weld retired to a quiet life. In his later years, he worked as a schoolteacher.

He also fell in love, married and raised children. He would live long enough to see both his grandchildren and the end of slavery when the Thirteenth, Fourteenth and Fifteenth Constitutional Amendments were ratified.

After the Civil War, Weld was invited to give the commencement speech at Oberlin's graduation.

The Lane Rebels were also invited for an informal reunion.

Together they celebrated how things had changed.

Yet some things were still the same. Thome's daughter noticed, *"The Lane Seminary boys were proud of their chief."*[408]

Huntingdon Lyman also gave a speech reminiscing of all they had accomplished together. *"Fifty years ago when we left Lane Seminary, people told us we would regret it."*

"Now looking back, none of us would change a thing. We are humbled to have been a part of this great struggle. Today all of us are still faithful to our principles. Regardless of public opinion, we believe that love never fails."[409]

Weld died in 1895. He had never wanted any credit for his work. He never cared about publicity. He actually hid from honors or awards. The only thing he wanted was to please God's heart.

History quickly forgot about him.

Historians decided that the Civil War was caused by economics. It was just the agricultural South wrestling against the industrial North.

Textbooks were written, completely ignoring how the Lane Rebels had birthed college campus activism.

But all that would change.

In 1889, a man named Gilbert Barnes was born. He would write one book that would change the history books of America.

For twenty-five years, Barnes taught Economics at Ohio Wesleyan University.

One day, one of his students said that no revival had ever had any lasting influence on social issues in American history.[410]

That statement really bothered Barnes. It made him go to the library and began researching.

While living in Ohio, he had heard one of his friends talk about their grandfather, James Birney, who had freed his slaves because of someone named Theodore Weld.

Barnes dug deeper and found the grandchildren of Theodore Weld. They were more than happy to tell him all about how their grandfather Weld had turned the tide of American history.

They had saved a treasure trove of historical documents from years before the Civil War. In their attic was a chest with stacks and stacks of handwritten letters between Weld, Charles Finney, Lewis Tappan and the Lane Rebels.

Barnes also found eyewitness accounts published by everyone from Asa Mahan to Henry Stanton.

Piece by piece, Barnes put together this remarkable story.

Publishing his stunning book in the 1930's, *The Antislavery Impulse,* he carefully traced how a revival really had sparked a revolt.

He found how the American people had sent so many thousands of abolition petitions to Congress that when he visited the records rooms of Congress in the 1930's, he discovered one of the staff members had been using the abolition petitions as fuel for his fire, yet still there were huge piles and piles of petitions left all over the place.

His research had a powerful effect on America, coming just in time for the Civil Rights movement.

Most importantly, it would show people how much one person could accomplish when following the leading of the Holy Spirit. Never again could historians deny that Christian people really can change society and politics.

This might be a photo of the soldiers from Wisconsin with the
teen girl they rescued. The girl's name is unknown. But one of
the soldiers in the photo was from that Wisconsin Army
Regiment known to be abolitionists. It would make sense that
they would want to take a photo to brag about protecting her.

Bibliography

Abzug, Robert. *Passionate Liberator: Theodore Dwight Weld.* New York: Oxford University Press, 1980.

Adams, John, and Abigail Adams. *Familiar Letters of John Adams and His Wife Abigail Adams.* Edited by Charles F. Adams. Boston: Houghton, Mifflin,   1875. Kindle Edition.

Adams, John Quincy. *Address of John Quincy Adams, to His Constituents of the Twelfth Congressional District, at Braintree, September 17th, 1842.* Boston: J.H. Eastburn, Printer, 1842. Google Books.

Adams, John Quincy. *Argument of John Quincy Adams, before the Supreme Court of the United States. Delivered on the 24th of February and 1st of March, 1841.* New York: S.W. Benedict, 1841. Google Books.

Adams, John, Abigail Adams, and Charles F. Adams. *Familiar Letters of John Adams and His Wife Abigail Adams.* Boston: Houghton, Mifflin, 1875. Kindle Edition.

Adams, John Quincy, and Charles Francis Adams. *Memoir: Comprising Portions of His Diary from 1795 to 1848.* Vol. 1-12. Philadelphia: J. B. Lippincott, 1874. Google Books.

Adams, John Quincy, and John Greenleaf Whittier. *Letters from John Quincy Adams to His Constituents of the Twelfth Congressional District in Massachusetts. To Which Is Added His Speech in Congress, Delivered February 9, 1837.* Boston: Isaac Knapp, 1837. Google Books.

Adams, John Quincy. *Speech of John Quincy Adams in the House of Representatives On the State of the Nation Delivered May 25, 1830.* New York: H. R. Piercy, 1836. Archive.org.

Adams, John Quincy. *Speech of John Quincy Adams, upon the Right of the People, Men and Women, to Petition; Delivered in the House of Representatives, from the 16th of June to the 7th of July, 1838.* Washington: Printed by Gales and Seaton, 1838. Google Books.

Adams, John Quincy. *Speech of the Hon. John Quincy Adams, in the House of Representatives, on the State of the Nation: Delivered May 25, 1836.* New York: H.R. Piercy, 1836. Archive.org.

Adams, John Quincy, and Worthington Chauncey Ford. *Writings of John Quincy Adams.* Vol. 1-7. New York: Macmillan, 1913. Archive.org.

An Officer in the Late Army. *Complete History of the Marquis De Lafayette.* Hartford, CT: S. Andrus, 1846. Google Books.

Amazing Grace. Directed by Jean-Marc Vallee. Performed by Ioan Gruffudd, Emily Blunt, Rupert Friend & Benedict Cumberbatch. Beverly Hills, Calif: Twentieth Century Fox Home Entertainment, 2007. DVD.

Aptheker, Herbert. *American Negro Slave Revolts.* New York: International Publishers, 1993.

Aptheker, Herbert. *A Documentary History of the Negro People in the United States.* Vol. I. New York: Citadel Press, 1971.

Aristotle. *Aristotle's Politics.* Translated by Benjamin Jowett. Edited by H. W. Carless Davis. Oxford: Clarendon Press, 1908. Google Books.

Armistead, Wilson. *Anthony Benezet.* Philadelphia: Lippincott, 1859. Google Books.

Arnold, Isaac. *The History of Abraham Lincoln and the Overthrow of Slavery.* Clarke & Co., 1866.

Asbury, Francis. *Journal of the Rev. Francis Asbury Bishop of The Methodist Episcopal Church.* Vol. 1-3. New York: Bangs and Mason, 1821. Archive.org.

Baker, William. "William Wilberforce on the Idea of Negro Inferiority: (He Destroyed It In England)." *Journal of the History of Ideas* 31, no. 3 (Summer 1970): 433-40. JSTOR.

Bancroft, George. *History of the United States of America, from the Discovery of the Continent.* Vol. 1. New York: Appleton, 1888. Google Books.

Barnes, Gilbert Hobbs. *The Antislavery Impulse, 1830-1844,.* New York: Harcourt, Brace and World, 1964.

Beecher, Lyman. *Autobiography, Correspondence, Etc., of Lyman Beecher,* Ed. Charles Beecher. Vol. 1-2. New York: Harper & Brothers, 1866. Google Books.

Belmonte, Kevin Charles. *William Wilberforce: A Hero for Humanity.* Grand Rapids, MI: Zondervan, 2009.

Bennett, Lerone. *Before the Mayflower: A History of Black America.* Harmondsworth, Middlesex: Penguin Books, 1987.

Benton, Thomas, and John Rives, eds. *Abridgment of the Debates of Congress, from 1789 to 1856.* Vol. 1. New York: D. Appleton, 1857. Google Books.

Birney, Catherine H., and Angelina Emily Grimke. *The Grimke´ Sisters: Sarah and Angelina Grimke´, the First American Women Advocates of Abolition and Woman's Rights.* Boston: Lee and Shepard, 1885. Google Books.

Birney, William. *James G. Birney and His Times; the Genesis of the Republican Party with Some Account of Abolition Movements in the South before 1828.* New York: D. Appleton and, 1890. Google Books.

Blake, William O. *The History of Slavery and the Slave Trade, Ancient and Modern.* Columbus, OH: H. Miller, 1857. Archive.org.

Bleby, Henry. *Death Struggles of Slavery.* London: Hamilton, Adams, 1853. Google Books.

Bradley, James. "Brief Account of An Emancipated Slave Written By Himself At The Editors Request." *The Oasis* [Cincinnati] June 1834.

Brisbane, William Henry. *Slaveholding Examined in the Light of the Holy Bible.* Philadelphia: U.S. Job Print. Office, 1847. Google Books.

Brodie, Fawn McKay. *Thaddeus Stevens, Scourge of the South.* New York: Norton, 1959.

Bryant, Joshua. *Account of an Insurrection of the Negro Slaves in the Colony of Demerara Which Broke out on the 18th of August, 1823.* Georgetown, Demerara: Printed by A. Stevenson at the Guiana Chronicle Office, 1824. Google Books.

Campbell, Stanley W. *The Slave Catchers; Enforcement of the Fugitive Slave Law, 1850-1860.* Chapel Hill: University of North Carolina Press, 1970.

Carter, Clarence Edwin. *The Struggle in Congress over Abolition Petitions.* University of Wisconsin, 1906. Google Books.

Chase, Rev Benjamin. "1846 Interview With Ona Judge Staines." Www.ushistory.org. January 1, 1847.

Clarkson, Thomas. *History of the Rise, Progress, And Accomplishment of the Abolition of the African Slave Trade by The British Parliament.* Vol. I-II. Philadelphia: James P. Parke, 1808.

Coburn, Frank. *Battle of April 19, 1775.* Lexington: Coburn, 1912. Google Books.

Child, Lydia Maria. *Memoir of Benjamin Lay.* New-York: American AntiSlavery Society, 1842. *Google Books.*

Cochran, Thomas. "William Wilberforce and the Abolition of the British Slave Trade: A Study of a Reform Movement." Master's thesis, Eastern Illinois University, 1967.

The Code of 1650, Being a Compilation of the Earliest Laws and Orders of the General Court of Connecticut, Adopted by the Towns of Windsor, Hartford, and Wethersfield in 1638-9. Hartford: S. Andrus, 1822. Openlibrary.org.

The Colonial Laws of Massachusetts. Boston: Rockwell & Churchhill, 1889. Published by order of Boston City Council.

Coffin, Joshua. *An Account of Some of the Principal Slave Insurrections, and Others, Which Have Occurred, or Been Attempted, in the United States and Elsewhere, during the Last Two Centuries.* New York: American Antislavery Society, 1860. Archive.org.

Coffin, Levi. *Reminiscences of Levi Coffin, with the Stories of Numerous Fugitives Who Gained Their Freedom.* Cincinnati: Robert Clarke, 1880. Google Books.

Cousins, Norman. *In God We Trust: The Religious Beliefs and Ideas of the American Founding Fathers.* New York: Harper, 1958.

Crawford, Macdermot. *Madame De Lafayette and Her Family.* New York: James Pott, 1907. Archive.org.

Cushing, William. *The Case of Nathaniel Jennison From the Notes of Chief Justice William Cushing.* Boston: John Wilson, 1874. Google Books.

Davis, Varina. *Jefferson Davis, Ex-president of the Confederate States of America, a Memoir by His Wife.* New York: Belford Company, 1890. Archive.org.

Declaration of Independence of the United States of America in Congress July 4th, 1776. Kindle Edition.

Devens, R. M. "Struggle for the Right of Petition in Congress." *American Progress,* Chicago, IL: H. Heron, 1883. 252-62.

Dresser, Amos. *The Narrative of Amos Dresser.* New York: American Anti-Slavery Society, 1836. Google Books.

Dubois, William E. B. *The Suppression of the African Slave Trade to the United States of America 1638-1870.* Vol. I. New York: Harvard Historical Studies, 1896. Kindle Edition by Project Gutenberg.

Dumond, Dwight Lowell. *Antislavery; the Crusade for Freedom in America.* Ann Arbor: University of Michigan, 1961.

Dumond, Dwight L. "Race Prejudice and Abolition: New Views on the Anti-slavery Movement." *Michigan Alumnus Quarterly Review: Journal of University Perspectives* 41 (1934): 377-85. Google Books.

East of England Broadband Network Company. "The Abolition of Slavery Project." The Abolition of Slavery Project. 2009. http://abolition.e2bn.org/.

Ellis, J. M. *Oberlin and the American Conflict: An Address Delivered At Alumni Oberlin Reunion August 26, 1865.* Ohio: Oberlin, 1865. Archive.org.

Fairchild, James Harris. *The Underground Railroad.* Cleveland, OH: Western Reserve Historical Society, 1895. Google Books.

Falkner, Leonard. *The President Who Wouldn't Retire.* New York: Coward-McCann, 1967.

Farrand, Max. *The Records of the Federal Convention of 1787.* Vol. 1-3. New Haven: Yale University Press, 1911. Google Books.

Federer, William J. *America's God and Country. Encyclopedia of Quotations.* St. Louis, MO: Amerisearch, 2000.

Fiske, John. *The Discovery of America: With Some Account of Ancient America and the Spanish Conquest.* Vol. II. Boston: Houghton, Mifflin, 1892.

Fithian, Philip Vickers. *Philip Vickers Fithian, Journal and Letters, 1767-1774.* Edited by John Rogers Williams. Princeton, NJ: University Library, 1900. Archive.org.

Fletcher, Robert Samuel. *A History of Oberlin College from Its Foundation through the Civil War.* Oberlin, O.: Oberlin College, 1943.

Finney, Charles G. *Charles G. Finney: An Autobiography.* Old Tappan, NJ: Revell, 1908. Google Books.

Franklin, Benjamin. *The Complete Works of Benjamin Franklin: Including His Private as Well as His Official and Scientific Correspondence.* Edited by John Bigelow. Vol. 1-10. New York: G.P. Putnam's Sons, 1887. Google Books.

Franklin, Benjamin. *Memoirs of the Life and Writings of Benjamin Franklin.* Edited by William Temple Franklin. Vol. 1. London: Printed for Henry Colburn, 1818. Note: This book written by Franklin's grandson.

Franklin, Benjamin. *The Works of Benjamin Franklin.* Edited by Jared Sparks. Vol. 5. Boston: Tappan & Whittemore, 1837. Google Books.

Franklin, John Hope, and Loren Schweninger. *Runaway Slaves: Rebels on the Plantation.* New York: Oxford University Press, 1999.

Garrison, William Lloyd. *The Letters of William Lloyd Garrison.* Edited by Walter Merrill and Louis Ruchames. Cambridge: Harvard University Press, 1976.

Garrison, Wendell Phillips, and Francis Jackson Garrison. *William Lloyd Garrison 1805-1879; Story of His Life As Told By His Children.* Vol. 1-4. New York: Houghton, Mifflin & Co, 1885. Archive.org.

Gates, Henry Louis. *Life upon These Shores: Looking at African American History, 1513-2008.* New York: Alfred A. Knopf, 2011.

Geneseo Centennial History: 1836-1936. Kewanee, IL: Printed by the Star-Courier, 1936. Archive.org.

Giddings, Joshua R. *History of the Rebellion: Its Authors and Causes.* New York: Follett, Foster &, 1864. Archive.org.

Greene, Lorenzo J. *The Negro in Colonial New England.* New York: Atheneum, 1968.

Grimke, Angelina Emily, and Catharine Esther Beecher. *Letters to Catherine Esther Beecher in Reply to an Essay on Slavery and Abolitionism Addressed to A.E. Grimké.* Boston: Printed by Isaac Knapp, 1838. Google Books.

Grimke, Angelina Emily. *Appeal to the Christian Women of the South.* [New York]: American Anti-Slavery Society, 1836. Google Books.

Grimké, Sarah M. *Letter to the Clergy of the Southern States.* New York: American Anti-Slavery Society, 1836. Print.

Goodell, Abner C., Jr. "John Saffin and His Slave Adam." In *Publications of the Colonial Society of Massachusetts: 1892-1894*, 85-112. Vol. 1. Boston: Self-Published, 1895. Google Books.

Goodell, William. *Slavery and Anti-Slavery; a History of the Great Struggle*. New York: William Goodell, 1855. Google Books.

Grimsted, David. *American Mobbing: 1828-1861: Toward Civil War*. New York: Oxford Univ. Press, 1998.

Hague, William. *William Wilberforce: The Life of the Great Anti-Slave Trade Campaigner*. Orlando: Harcourt, 2007.

Hammond, James Henry. *Remarks of Mr. Hammond, of South Carolina, on the Question of Receiving Petitions for the Abolition of Slavery*. Washington City: D. Green, 1836. Google Books.

Hanke, Lewis. *The Spanish Struggle for Justice in the Conquest of America*. Philadelphia: University of Pennsylvania Press, 1949. Archive.org.

Hardin, William. *Litigating the Lash*. Master's thesis, Vanderbilt University, 2013.

Harford, John S. *Recollections of William Wilberforce*. London: Longman, Green, Roberts, 1865. Google Books.

Harrold, Stanley. "On the Borders of Slavery and Race: Charles T. Torrey and the Underground Railroad." *Journal of the Early Republic* 20, no. 2 (Summer 2000): 273-92. JSTOR.

Hart, Albert Bushnell. *Slavery and Abolition, 1831-1841*. New York: Harper & Bros., 1906. Archive.org

Henry, Stuart. "The Lane Rebels: A Twentieth Century Look." *Journal of Presbyterian History* 49, no. 1 (Spring 1971): 1-14. JSTOR.

Hepburn, John. *American Defense of the Christian Golden Rule: Essay to Prove Unlawfulness of Making Slaves of Men*. New York: Self-published, 1715. Evans Early American Imprint Collection. http:/quod.lib.umich.edu/

Hinde, Thomas. "Speech of Patrick Henry." Edited by J. M. Peck. *Baptist Memorial and Monthly Record* 4, no. 5 (May 1845): 129-32. Google Books.

Hirschfeld, Fritz. *George Washington and Slavery: A Documentary Portrayal*. Columbia: University of Missouri Press, 1997.

Hoare, Prince. *Memoirs of Granville Sharp*. London: H. Colburn, 1820. Archive.org.

Hoffman, Michael A. *They Were White and They Were Slaves; the Untold History of the Enslavement of Whites in Early America*. Dresden, N.Y: Wiswell Ruffin House, 1992.

Hutchins, Frank, and Cortelle Hutchins. *Sword of Liberty; the Story of Two Revolutions*. New York: Century, 1921. Google Books.

Ingham, John N. "Lewis Tappan." In *Biographical Dictionary of American Business Leaders*, 1424-426. Westport, CT: Greenwood Press, 1983.

Jackson, Maurice. *Let This Voice Be Heard: Anthony Benezet, Father of Atlantic Abolitionism*. Philadelphia: University of Pennsylvania Press, 2009. Kindle Edition.

Jamaica Information Service. "Samuel Sharpe: Jamaican National Hero." Government of Jamaica. 2014.

Jefferson, Thomas. *The Writings of Thomas Jefferson*. Edited by H. A. Washington. Vol. 3. Washington, D.C.: Taylor & Maury, 1853. Google Books.

Johnson, William. *Abraham Lincoln the Christian*. New York: Eaton & Mains, 1913. Google Books.

Johnson, William J. *George Washington, the Christian*. New York: Abingdon Press, 1919. Google Books.

Jordan, Don, and Michael Walsh. *White Cargo: The Forgotten History of Britain's White Slaves in America*. Washington Square, NY: New York University Press, 2008.

Journals of the Continental Congress 1774-1789. Vol. 1. Washington: U.S. Government Printing Office, 1904. Google Books.

Justice, Hilda. *Life and Ancestry of Warner Mifflin*, Philadelphia: Ferris & Leach, 1905. Archive.org.

Kaminski, John P. *A Necessary Evil? Slavery and the Debate over the Constitution*. Madison, WI: Madison House, 1995.

Keitt, Lawrence M. *Speech of Hon. Lawrence M. Keitt, of South Carolina, on Slavery, and the Resources of the South: Delivered in the House of Representatives, January 15, 1857*. Washington: Printed at the Office of the Congressional Globe, 1857. Archive.org.

Kennedy, Melvin. *Lafayette and Slavery, from His Letters to Thomas Clarkson and Granville Sharp*. Easton, PA: American Friends of Lafayette, 1950.

Koger, Larry. *Black Slaveowners: Free Black Slave Masters in South Carolina, 1790-1860*. Columbia, SC: University of South Carolina Press, 1995.

Lafayette, Marquis De. *Lafayette in the Age of the American Revolution: Selected Letters and Papers, 1776-1790*. Edited by Stanley Idzerda. Vol. 1-5. Ithaca, NY: Cornell University Press, 1983.

Lafayette, Marquis De. *Memoirs, Correspondence and Manuscripts of General Lafayette Published By His Family*. Translated by William Duer. Vol. 1-2. New York: Saunders, 1837. Google Books.

Lawson-Jones, LaTonya. "History of the Carter Family." Nomini Hall Slave Legacy. 2012.

Lay, Benjamin. *All Slave-keepers That Keep the Innocent in Bondage (are) Apostates*. Philadelphia: Printed by Benjamin Franklin, 1737. Archive.org.

Lean, Garth. *God's Politician: William Wilberforce's Struggle*. Colorado Springs, CO: Helmers & Howard, 1987.

Lee, Jason K. *The Theology of John Smyth: Puritan, Separatist, Baptist, Mennonite*. Macon, GA: Mercer University Press, 2003.

Lerner, Gerda. *The Grimké Sisters from South Carolina: Rebels against Slavery*. Boston: Houghton Mifflin, 1967.

Lesick, Lawrence Thomas. *The Lane Rebels: Evangelicalism and Antislavery in Antebellum America*. Metuchen, NJ: Scarecrow, 1980.

Levasseur, Auguste. *Lafayette in America 1824-1825: Written By His Secretary*. Vol. 1-2. New York: White & Gallagher, 1829. Google Books.

Levine, Bruce C. *The Fall of the House of Dixie: The Civil War and the Social Revolution That Transformed the South*. New York: Random House, 2013. Kindle Edition.

Levy, Andrew. "The Anti-Jefferson: Why Robert Carter III Freed His Slaves (And Why We Couldn't Care Less)." *The American Scholar* 70, no. 2 (Spring 2001): 15-35. JSTOR.

Levy, Andrew. *The First Emancipator: The Forgotten Story of Robert Carter, the Founding Father Who Freed His Slaves*. New York: Random House, 2005.

Lilly, Lambert. *The Adventures of Captain John Smith, the Founder of the Colony of Virginia*. New York: D. Appleton &, 1842. Google Books.

Lincoln, Abraham. *The Emancipation Proclamation*. Washington, D.C.: U.S. Gov. Printing Office, 1862. Kindle Edition.

Livermore, George. *Historical Research*. New York: J. Wilson and Son, 1862. *Google Books*.

Lloyd, Christopher. *The Navy and the Slave Trade: The Suppression of the African Slave Trade in the Nineteenth Century*. London: Cass, 1968.

Locke, Mary Stoughton. *Anti-Slavery in America: From the Introduction of African Slaves to the Prohibition of the Slave Trade, 1619-1808*. Gloucester, MA: Peter Smith, 1965.

Losada, Angel. "Bartolome De Las Casas Champion of Indian Rights in 16th Century." The Courier, June 1975, 4-10. doi:Printed by UNESCO: The United Nations Educational, Scientific and Cultural Organization.

Loosemore, Jo. "Sailing Against Slavery." BBC. August 7, 2008.

Lovejoy, Joseph C. *Memoir of Rev. Charles T. Torrey Who Died in the Penitentiary of Maryland: Where He Was Confined for Showing Mercy to the Poor*. Boston: J.P. Jewett &, 1847. Google Books.

Lovejoy, Owen, William F. Moore, and Jane Ann. Moore. *His Brother's Blood: Speeches and Writings, 1838-64*. Urbana: University of Illinois Press, 2004.

Lundy, Benjamin, and Thomas Earle. *The Life, Travels and Opinions of Benjamin Lundy*. Philadelphia: W.D. Parrish, 1847. Google Books.

Lyman, Huntingdon. "Lane Seminary Rebels." *Oberlin Jubilee* 1883: 60-69.

Mack, Ebenezer, and Jules Cloquet. *Life of Gilbert Motier De Lafayette: A Marquis of France.* New York: Mack, Andrus, & Woodruff, 1841. Google News.

Madison, James. *Writings of James Madison: Journal of the Constitutional Convention.* Edited by Gaillard Hunt. Vol. 3. New York: G. P. Putnam, 1902. Google Books.

Mahan, Asa. *Intellectual, Moral, and Spiritual Autobiography.* London: Self-Published, 1882. Google Books.

Malone, Dumas. *Jefferson and the Rights of Man.* Vol. 1-2. Boston: Little, Brown &, 1951. Archive.org.

Marshall, Peter, and David Manuel. *The Light and the Glory.* Old Tappan, NJ: Revell, 1977.

Marshall, Peter, and David Manuel. *From Sea to Shining Sea, 1787-1837.* Grand Rapids, MI: Revell, 2009.

Marshall, Peter, and David Manuel. *Sounding Forth the Trumpet.* Grand Rapids, MI: Fleming H. Revell, 1997.

Marshall, Thomas Francis, and W. L. Barre. *Speeches and Writings of Hon. Thomas F. Marshall.* Cincinnati: Applegate &, 1858. Google Books.

McDowell, Michael. "Warner Mifflin: A Founding Father of Abolition-ism." *Quaker Hill Quill* 2, no. 3 (Summer 2013): 1-4.

Miller, William Lee. *Arguing about Slavery: The Great Battle in the United States Congress. New York: A.A. Knopf, 1996.*

Moore, George. *Notes on the History of Slavery in Massachusetts.* New York: John F. Trow &, 1866. Archive.org.

Morton, Louis. *Robert Carter of Nomini Hall, a Virginia Tobacco Planter of the Eighteenth Century.* Williamsburg, VA: Colonial Williamsburg, 1941.

Maxwell, John Francis. *Slavery and the Catholic Church: The History of Catholic Teaching concerning the Moral Legitimacy of the Institution of Slavery. Forward by Lord Wilberforce.* Chichester: Anti-Slavery Society for the Protection of Human Rights, 1975.

Marley, David. *Wars of the Americas: A Chronology of Armed Conflict in the New World, 1492 to the Present.* Santa Barbara, CA: ABC-CLIO, 1998.

Metaxas, Eric. *Amazing Grace: William Wilberforce and the Heroic Campaign to End Slavery.* New York, NY: Harper San Francisco, 2007.

National Archives. "To George Washington From Warner Mifflin: November 23, 1792." Founders Online. www.founders.archives.gov.

Nell, William C. *The Colored Patriots of the American Revolution.* Boston: R.F. Wallcut, 1855. Archive.org.

Niles, Hezekiah. *Principles and Acts of the Revolution in America:* Baltimore: Printed by W.O. Niles, 1822. Google Books.

"The Oberlin-Wellington Rescue 1858." Oberlin Heritage Center. Ed. Liz Schultz, Roland Baumann, Gary Kornblith, and Mary Moroney. N.p., 6 Jan. 2009. Web.

"Our History: William Wilberforce & Family." Westminister Abbey. 2015. www.westminister-abbey.org.

Peck, J. M. *"Speech of Patrick Henry."* In Baptist Memorial and Monthly Record, 129-33. Vol. 4. New York: John R. Bigelow, 1845.

Parker, Theodore. *The Trial of Theodore Parker: For the "Misdemeanor" of a Speech in Faneuil Hall against Kidnapping, before the Circuit Court of the United States, at Boston, April 3, 1855.* Boston: Published for the Author, 1855. Google Books.

The Parliamentary Debates: Official Report. Vol. VII. London: Thomas C. Hansard, 1812. Google Books.

Pelteret, David. *Slave Raiding and Slave Trading in Early England.* Cambridge: Cambridge University Press, 1981.

Pettit, Eber M. *Sketches in the History of the Underground Railroad.* Fredonia, NY: W. McKinstry & Son, 1879. Archive.org.

Pritchard, James. "Into the Fiery Furnace" Anti-Slavery Prisoners in the Kentucky State Penitentiary 1844-1870." Kentucky's Underground Railroad. 2006.

Pickett, Margaret F., and Dwayne W. Pickett. *The European Struggle to Settle North America: Colonizing Attempts by England, France and Spain, 1521-1608.* Jefferson, NC: McFarland &, 2011.

Piper, Emilie, and David Levinson. *One Minute a Free Woman: Elizabeth Freeman and the Struggle for Freedom.* Salisbury, CT: Published by Upper Housatonic Valley National Heritage Area, 2010.

Plato. *Plato's Best Thoughts.* Compiled by Benjamin Jowett. Translated by C. H. A. Bulkley. New York: Charles Scribner's Sons, 1883. Google Books.

Pollock, John Charles. *William Wilberforce.* Oxford: Lion, 1986.

Pope Leo XIII. "Pope Leo XIII On the Abolition of Slavery May 5, 1888."*Papal Encyclicals Online.* N.p., n.d. Web. <www.papalencyclicals.net>.

Proceedings of the Convention of Delegates of the Colony of Virginia Held March 20, 1775. Richmond: Ritchie, Trueheart Duval, 1816. Google Books.

Quincy, Josiah. *Memoir of the Life of John Quincy Adams.* Boston: Crosby, Nichols, Lee, 1860. Google Books.

Quarles, Benjamin. *Black Abolitionists.* New York: Oxford University Press, 1969.

Ramsay, James. *An Essay on the Treatment and Conversion of African Slaves in the British Sugar Colonies.* London: Printed and Sold by James Phillips, 1784. Archive.org.

Reid-Salmon, Delroy A. *Burning for Freedom: A Theology of the Black Atlantic Struggle for Liberation.* Kingston: Ian Randle Publishers, 2012.

Richards, Leonard L. *Gentlemen of Property and Standing; Anti-Abolition Mobs in Jacksonian America*. New York: Oxford University Press, 1970.

Rodriguez, Junius. *Chronology of World Slavery*. Santa Barbara, CA: ABC-CLIO, 1999.

Rushing, S. Kittrell. *Memory and Myth: The Civil War in Fiction and Film from Uncle Tom's Cabin to Cold Mountain*. West Lafayette, IN: Purdue University Press, 2007.

Russell, John H. *The Free Negro in Virginia: 1619-1865*. Baltimore: Johns Hopkins Press, 1913. Archive.org.

Russell, John H., PhD. "Colored Freemen As Slave Owners in Virginia." Journal of Negro History I, no. 3, 232-42. Archive.org.

Sandiford, Ralph. *A Brief Examination of the Practice of the Times*. Philadelphia: Printed by Ben Franklin, 1729. Evans Early American Imprint Collection. http://name.umdl.umich.edu/

Sedgewick, Catharine. "Slavery in New England." In *Bentley's Miscellany*, 417-24. Vol. 34. London: Richard Bentley, 1853. Google Books.

Sewall, Samuel. *The Selling of Joseph; a Memorial*. Boston: Bartholomew Green and John Allen, 1700. Archive.org.

Shurtleff M.D., Nathaniel, ed. *Records of the Governor and Company of the Massachusetts Bay*. Vol. 1-2. Boston: Printed by Order of the Massachusetts Legislature, 1853. Google Books.

Siebert, Wilbur Henry. *The Underground Railroad from Slavery to Freedom*. New York: Macmillan, 1898. Archive.org.

Seward, William H. *Life and Public Services of President John Quincy Adams*. New York: Derby, Miller, &, 1849. Project Gutenberg Kindle Edition.

Shipherd, Jacob R., Ralph Plumb, and Henry E. Peck. *History of the Oberlin-Wellington Rescue*. Boston: J.P. Jewett and, 1859. Google Books.

Sluiter, Engel. "New Light on The Twenty and Odd Negroes Arriving in Virginia August 1619." William and Mary Quarterly, 3rd ser., 54, no. 2 (April 1997): 395-98. JSTOR.

Smith, Gerrit. *Speeches of Gerrit Smith in Congress*. New-York: Mason Bros., 1856. Google Books.

Smith, John. *Capt. John Smith: President of Virginia, and Admiral of New England: Works, 1608-1631*. Edited by Edward Arber. Birmingham: English Scholar's Library, 1884. Archive.org.

Society of Friends. "Early Antislavery Advocates." In *The Friend*. Vol. 29. Philadelphia: Robb, Pile & Melroy, 1856. Google Books.

Sparks, Jared. *The Life of Benjamin Franklin: Containing the Autobiography*. Boston: Tappan and Dennet, 1844. Google Books.

Spector, Robert M. "The Quock Walker Cases (1781-83): Slavery, Its Abolition, and Negro Citizenship in Early Massachusetts." *The Journal of Negro History* 53, no. 1 (January 1968): 12-32. JSTOR.

Stanton, Henry B. *Random Recollections.* New York: Harper & Brothers, 1887. Google Books.

Stowe, Harriet Beecher. *A Key to Uncle Tom's Cabin; Presenting the Original Facts and Documents upon Which the Story Is Founded.* London: Sampson, Low, Son, 1853. Google Books.

Substance of the Debates on a Resolution for Abolishing the Slave Trade: Which Was Moved in the House of Commons on the 10th June, 1806, and in the House of Lords on the 24th June, 1806. London: Printed and Sold by Phillips and Fardon, George Yard, Lombard Street, 1806. Google Books.

Sumner, Charles. "Lafayette: The Faithful One." Lecture. Boston: Wright & Potter, 1870.

Tappan, Lewis. *The Life of Arthur Tappan.* New York: Hurd and Houghton, 1870. Google Books.

Thiers, Adolphe. *The History of the French Revolution.* Translated by J. Dixon. Vol. 1. London: G. Vickers, 1845. Google Books.

Thomas, Benjamin Platt. *Theodore Weld, Crusader for Freedom.* New Brunswick: Rutgers UP, 1973.

Thomas, Hugh. *The Slave Trade: The Story of the Atlantic Slave Trade, 1440-1870.* New York: Simon & Schuster, 1997.

Thomas, Hugh. *Rivers of Gold: The Rise of the Spanish Empire, from Columbus to Magellan.* New York: Random House, 2003.

Thome, James A., Samuel H. Cox, and Henry Stanton. *Debate at the Lane Seminary, Cincinnati: Speech of James A. Thome, of Kentucky, Delivered at the Annual Meeting of the American Anti-Slavery Society, May 6, 1834.* Boston. Published by Garrison & Knapp, No. 11, Merchants' Hall., 1834. Google Books.

Thornton, John Wingate. *The Pulpit of the American Revolution,* Gould and Lincoln, 1860. Google Books.

Tomkins, Stephen. *William Wilberforce: A Biography.* Grand Rapids, MI: William B. Eerdmans Pub., 2007.

Unger, Harlow G. *Lafayette.* New York: John Wiley & Sons, 2002.

United States Congress, House of Representatives. *Congressional Globe.* Library of Congress. www.memory.loc.gov.

United States Congress House of Representatives. *Addresses in the Congress and Funeral Solemnities on the Death of John Quincy Adams, February 23, 1848.* Washington, D.C.: G. S. Gideon, 1848. Google Books.

U. S. Supreme Court. *Reports of Decisions in the Supreme Court of the United States.* Edited by Benjamin Robbins Curtis. Vol. 15. Boston: Little, Brown, 1855. Google Books.

U.S. Continental Congress. *Journals of the American Congress from 1774-1788: Aug 1, 1778 to March 30, 1782.* Vol. 1-4. Washington, D.C.: Way and Gideon, 1823. Google Books.

United States Congress House of Representatives. *Annals of Congress.* Library of Congress. www.memory.loc.gov.

Urbainczyk, Theresa. *Slave Revolts in Antiquity.* Berkeley: University of California Press, 2008.

Vaux, Roberts. *Memoir of Benjamin Lay: Compiled from Various Sources.* Philadelphia: Conrad, 1815. Archive.org.

Vaver, Anthony. *Bound with an Iron Chain: The Untold Story of How the British Transported 50,000 Convicts to Colonial America.* Westborough, MA: Pickpocket Publishing, 2011. Kindle Edition.

Waldman, Steven. *Founding Faith: Providence, Politics, and the Birth of Religious Freedom in America.* New York: Random House, 2008.

Washington, George. *Writings of George Washington.* Edited by Jared Sparks. Vol. 1. Boston: Russell, 1837. Google Books.

Washington, George. *The Writings of George Washington.* Edited by Jared Sparks. Vol. 5. New York: Harper & Brothers, 1847. Google Books.

Weld, Theodore. *Does the Bible Sanction Slavery?* New York: American Antislavery Society, 1838. Google Books.

Weld, Theodore Dwight, Angelina Emily Grimke, and Sarah Moore Grimke. *Letters of Theodore Dwight Weld, Angelina Grimke and Sarah Grimke, 1822-1844.* Ed. Gilbert Hobbs Barnes and Dwight Lowell Dumond. Vol. 1-2. New York: D. Appelton-Century, 1934.

Weld, Theodore Dwight. *The Bible Against Slavery.* New York, NY: American Anti-Slavery Society, 1838. Google Books.

Weld, Theodore. *Does the Bible Sanction Slavery?* New York: American Anti-slavery Society, 1838. Google Books.

Weld, Theodore Dwight, Horace Moulton, Sarah Moore Grimké, Angelina Grimke, and American Antislavery Society. *American Slavery as It Is Testimony of a Thousand Witnesses.* New York: American Anti-Slavery Society, 1839. Kindle Edition.

Weld, Theodore Dwight. *In Memory, Angelina Grimke Weld: 1805-1879.* Boston: Press of G.H. Ellis, 1880. Google Books.

Weld, Theodore. *A Statement of the Reasons Which Induced the Students of Lane Seminary to Dissolve Their Connection with That Institution: December 15, 1834.* Cincinnati: Lane Rebels, 1834. Archive.org

Wheelan, Joseph. *Mr. Adams's Last Crusade: John Quincy Adams's Extraordinary Post-Presidential Life in Congress.* New York: PublicAffairs, 2008.

Wiencek, Henry. *An Imperfect God: George Washington, His Slaves, and the Creation of America.* New York: Farrar, Straus and Giroux, 2003.

Wilson, Henry. *History of the Rise and Fall of the Slave Power in America.* Vol. 1-3. Boston: J.R. Osgood, 1872. Google Books.

Wilberforce, Robert Isaac, and Samuel Wilberforce. *The Life of William Wilberforce.* Vol. 1-5. London: J. Murray, 1838. Google Books.

Wilberforce, William. *The Correspondence of William Wilberforce.* Vol. 1-2. London: John Murray, 1840. Google Books.

Wilberforce, William. *A Letter on the Abolition of the Slave Trade.* London: Printed by Luke Hansard & Sons for T. Cadell and W. Davies, and J. Hatchard, 1807. Google Books.

Wilberforce, William. *A Practical View of the Prevailing Religious System of Professed Christians, Contrasted with Real Christianity.* Dublin: Robert Dapper, 1797. Kindle Edition.

Wilberforce, William. *Private Papers of William Wilberforce.* Compiled by Anna Maria Wilberforce. London: T. Fisher, 1897. Archive.org.

Wijngaards, John. "Discussion on the Popes and Slavery." *Wijngaards Institute for Catholic Research.* N.p., 1998. <www.womenpriests.org>.

Winthrop, John. *The History of New England from 1630 to 1649.* Ed. James Savage. Vol. 1-2. Boston: Phelps and Farnham, 1825. Archive.org.

Wilson, Henry. *History of the Rise and Fall of the Slave Power in America.* Vol. 1-3. Boston: J.R. Osgood and, 1872. Archive.org.

Wilds, Mary. Mumbet: *The Life and Times of Elizabeth Freeman: The True Story of a Slave Who Won Her Freedom.* Greensboro, NC: Avisson Press, 1999.

Winthrop, John. *The History of New England from 1630 to 1649.* Ed. James Savage. Vol. 1-2. Boston: Phelps and Farnham, 1825. Archive.org.

Winthrop, John. *History of New England.* Edited by James Hosmer. Vol. 1. New York: Charles Schribner, 1908. Google Books.

Wisconsin Historical Soceity. *Proceedings of the Forty-Second Annual Meeting of Wisconsin Srate Historical Society.* Madison, Wisconsin: Democrat Printing, 1895. Google Books.

Wyatt-Brown, Bertram. *Lewis Tappan and the Evangelical War against Slavery.* Cleveland: Press of Case Western Reserve U, 1969.

Zwelling, Shomer S. "Robert Carter's Journey From Colonial Patriarch to New Nation Mystic." *American Quarterly* 38, no. 4 (Autumn 1986): 613-36. JSTOR.

Bible Quotes

Note: Bible verses in this book are from the KJV unless otherwise stated.

CEB: Scripture quotations marked CEB are from the Holy Bible, Common English Bible. Copyright © 2011 by Common English Bible.

ESV: Scripture quotations marked ESV are from the ESV® Bible (The Holy Bible, English Standard Version®), copyright© 2001 by Crossway Bibles, a publishing ministry of Good News Publishers. Used by permission. All rights reserved.

BSB: Scripture quotations marked BSB are from The Holy Bible, Berean Study Bible, BSB, Copyright ©2016, 2020 by Bible Hub. Used by Permission. All Rights Reserved Worldwide.

GW: Scripture quotations marked GW are from the Holy Bible, God's Word Translation. Copyright © 1995, 2003, 2013, 2014, 2019, 2020 by God's Word to the Nations Mission Society. Used by Permission. All rights reserved.

KJV: Scripture quotations marked KJV are from the Holy Bible, King James Version, which is in the public domain.

NKJV: Scripture quotations marked NKJV are from the Holy Bible, New King James Version®. Copyright © 1982 by Thomas Nelson. Used by permission. All rights reserved.

NIV: Scripture quotations marked NIV taken from the Holy Bible, New International Version ®, NIV ® Copyright © 1973, 1978, 1984, 2011 by Biblica, Inc. Used with permission. All rights reserved worldwide.

NLT: Scripture quotations marked NLT are from the Holy Bible, New Living Translation, copyright © 1996, 2004, 2015 by Tyndale House Foundation. Used by permission of Tyndale House Publishers, Inc., Carol Stream, Illinois 60188. All rights reserved.

References

1. ^ Wilberforce, Robert, and Samuel Wilberforce. *Life*, Vol 1. p. 149.

2. ^ Belmonte, Kevin. *William Wilberforce*, 61 & 223 & Metaxas, Eric. *Amazing Grace*, 29 & Lean, Garth. *God's Politician*, 22 & Pollock, John. *William Wilberforce*, 24.

3. ^ Wilberforce, William. *Private Papers*, 13.

4. ^ Wilberforce, William. *Correspondence*, Vol 1, 55-59.

5. ^ Wilberforce, Robert, and Samuel Wilberforce. *Life of Wilberforce*, Vol 1. 187- 188.

6. ^ Gates, Henry Louis. *Life upon These Shores*, 6-7.

7. ^ Ibid, 8-9.

8. ^ Gates, Henry Louis. *Life upon These Shores*, 10-11.

9. ^ Winthrop, John. *The History of New England from 1630 to 1649*, Vol 2. 327.

10. ^ Blake, William O. *History*, 158 & Hague, *Wilberforce*, 116.

11. ^ Blake, William O. *History*, 106.

12. ^ Ibid, 165.

13. ^ The actual copy of Granville Sharp's book that was owned by John Adams is still in the historical collection of the Boston public library. The book title was *The Laws of God Compared with the Claims of Slaveholders*.

14. ^ Blake, William O. History, 106 & Lloyd, *The Navy and The Slave Trade*, 10.

15. ^ Dubois, William E. B. *Suppression*, Chapter 9.

16. ^ Goodell, William. *Slavery*, 97.

17. ^ Hirschfeld, Fritz. *George Washington and Slavery*, 130-132.

18. ^ Ibid, 134-135.

19. ^ Wilberforce, Robert and Samuel Wilberforce. *Life*, Vol 1, 147.

20. ^ Ibid, Vol 1, 149-150.

21. ^ Blake, William O. *History*, 195.

22. ^ Harford, John S. *Recollections*, 139.

23. ^ Wilberforce, Robert, et al, *Life*, Vol 1. 153.

24. ^ Blake, William O. *History*, 191-214. The note on England's trade increasing by 83% after abolishing slavery comes from the records of the House of Commons debates referenced in Lloyd, Christopher. *The Navy and the Slave Trade*, 147.

25. ^ Cochran, Thomas. *William Wilberforce*, 25 & Hague, *Wilberforce*, 183 & Pollock, John Charles. *William Wilberforce*, 89.

26. ^ Blake, William O. *History*, 211-212 & Hague, *Wilberforce*, 197-198.

27. ^ Blake, William O. *History*, 212 & 230, & Hague, *Wilberforce*, 197.

28. ^ Blake, William O. *History*, 213-215.

29. ^ Parker, Theodore. *Trial*, 39.

30. ^ Blake, William O. *History*, 217 & Hague, *Wilberforce*, 197.

31. ^ Blake, William O. *History*, 221-224.

32. ^ Thomas, Hugh. *The Slave Trade*, 31.

33. ^ Hague, *Wilberforce*, 228.

34. ^ Ibid, 263.

35. ^ Blake, William O. *History*, 225 & Hague, *Wilberforce*, 235.

36. ^ Garrison, et al. *Story of His Life Told By His Children*, Vol 3. 354.

37. ^ Hague, *Wilberforce*, 88 & Wilberforce, William. *Correspondence*, Vol 1. 56.

38. ^ Wilberforce, William. *Real Christianity*, Kindle Edition.

39. ^ *The Parliamentary Debates: Official Report.* Vol 7. p 32.

40. ^ Ibid, Vol 7. p 593. Wilberforce is referencing Exodus 21:16 when God gives the death penalty to slave traders.

41. ^ Johnson, William. *Abraham Lincoln the Christian*, 79 & Lloyd, *Navy*, 42 & 175.

42. ^ Johnson, *Lincoln the Christian*, 79 & Lloyd, *Navy*, 42, 58-59, 175.

43. ^ Lloyd, Christopher, *The Navy and the Slave Trade*, 46-47, 117, 175. (Lloyd's book quotes from official sources and is lower numbers than other references here so the author's opinion is that Lloyd's numbers are probably most accurate). Also see Loosemore, Jo. "Sailing Against Slavery." & Tomkins, Stephen. *William Wilberforce*, 222.

44. ^ Hauge, *Wilberforce,* 479.

45. ^ Bleby, Henry. *Death Struggles*, 291 & Tomkins, Stephen. *William Wilberforce*, 216.

46. ^ Wilberforce, et. al, *Life*, Vol 5. p. 370.

47. ^ "Our History: William Wilberforce & Family." Westminster Abbey.

48. ^ Moore, George. *Notes,* 74-76 & Rodriguez, Junius. *Chronology*, 413-414.

49. ^ Burling, William, *Some Observations*, Quoted in Ben Lay's book, 7-12. Also see Sewall, Samuel, *Selling of Joseph* & John Hepburn, *Defense of Golden Rule.*

50. ^ Sandiford, Ralph. *Brief Examination*, 35-36. He's quoting Isaiah 58 & 61.

51. ^ Child, Lydia Marie. *Memoir,* 25 & Vaux, Roberts. *Memoir,* 29 & 36-37.

52. ^ Vaux, Roberts. *Memoir,* 28, 36, & 40.

53. ^ Child, *Memoir,* 22 & Vaux, *Memoir,* 34.

54. ^ Society of Friends. "Early Antislavery Advocates." 212.

55. ^ Child, *Memoir,* 18 & Vaux, *Memoir,* 29.

56. ^ Lay, Benjamin, *Apostates*, Title page

57. ^ Ibid, 47-48.

58. ^ Ibid, 10, 25. (Lay is quoting Burling on p. 10).

59. ^ Ibid, 27, 46, 75, 144.

60. ^ Ibid, 41.

61. ^ Ibid, 114.

62. ^ Ibid, 26-28.

63. ^ Ibid, 95-97.

64. ^ Ibid, 95-97.

65. ^ Ibid. Psalms 50:16 (GW).

66. ^ Ibid, 30-31, 59, 85, 95-97, 165 & Revelations 2:5b (BSB)

67. ^ Lay, Apostates, *Apostates,* 30-31, 59, 85,165.

68. ^ Ibid, 67.

69. ^ Ibid, 4

70. ^ Ibid, 156.

71. ^ Ibid, 9, 179-185.

72. ^ Society of Friends, "Early Advocates." 220.

73. ^ Child, *Memoir,* 28 & Vaux, Roberts, *Memoir,* 42.

74. ^ Vaux, Roberts, *Memoir,* 51.

75. ^ Child, *Memoir of Benjamin Lay,* 19 & Vaux, Roberts, *Memoir,* 31.

76. ^ Child, 20 & Vaux 32.

77. ^ Jackson, *Voice,* 1.

78. ^ Franklin, Benjamin, *Works,* Vol 5. p. 128.

79. ^ Livermore, *Historical Research,* 16-17.

80. ^ Dubois, William. *Suppression,* Kindle Location 1340 & Livermore, *Historical Research,* 18.

81. ^ Livermore, *Historical Research,* 18.

82. ^ Declaration of Independence.

83. ^ *Journals of Continental Congress,* Vol 1. p. 76-77 & Livermore, *Historical Research,* 19.

84. ^ Hoarce, *Memoirs of Granville Sharp,* 118-119.

85. ^ Rodriguez, *Historical Chronology of World Slavery,* 20.

86. ^ Thomas, Hughes. *The Slave Trade,* 28.

87. ^ Aristotle, *Politics,* 32-34.

88. ^ Plato, *Plato's Best Thoughts*, 84.

89. ^ Urbainczyk, Theresa. *Slave Revolts in Antiquity*, 96-97.

90. ^ Cousins, Norman. *In God We Trust*, 226.

91. ^ Blake, William O. *History*, 43.

92. ^ Pope Leo XIII. "Pope Leo XIII On the Abolition of Slavery May 5, 1888."

93. ^ Mark 10:42b-43 (NLT) & 10:45 (KJV).

94. ^ Winthrop, John. *History of New England from 1630 to 1649*, Vol 2. 12-13.

95. ^ Gates, Henry Louis. *Life Upon These Shores*, 4-5.

96. ^ Rodriguez, Junius. *The Historical Chronology*, 408-411.

97. ^ Levine, Bruce C. *The Fall of the House of Dixie*, Kindle Location 6065.

98. ^ Hanke, Lewis, The Spanish Struggle, 17-18.

99. ^ Ibid.

100. ^ Smith, John. Capt. *John Smith*, XIV & 489-487 & 611.

101. ^ Russell, John H. *The Free Negro in Virginia*, 24.

102. ^ Ibid, 32.

103. ^ Greene, *Negro in Colonial New England*, 19.

104. ^ Koger, Larry. *Black Slaveowners*, 39.

105. ^ Ibid, 136 & 190.

106. ^ Ibid, 109.

107. ^ Ibid, 197.

108. ^ Greene, *Negro in Colonial New England*, 16 & Moore, George. *Notes*, 8-9.

109. ^ Moore, George. *Notes*, 10.

110. ^ Ibid.

111. ^ Winthrop, John. *History*, Vol 1. p. 260 & Moore, *Notes*, 4-6.

112. ^ Moore, George, *Notes*, 146-147.

113. ^ Cushing, William, *Case*, 5.

114. ^ Moore, George. *Notes*, 149-152. The story of the three men being rescued from the slave traders is from Nell, William. *Colored Patriots*, 59-60.

115. ^ Rodriguez, Junius. *Chronology*, 195. The story about the Vermont Judge demanding legal title from God is from Isaac Arnold's book, *Abraham Lincoln and the Overthrow of Slavery*, p. 28-29.

116. ^ Dubois, William E. B. *The Suppression*, Kindle Location 915 & Mary Locke, *AntiSlavery*, 14 & Nathaniel Shurtleff, *Records*, Vol 2, 168 & Henry Wilson, *History*, Vol 1, p. 6.

117. ^ Greene, *Negro in Colonial New England*, 18.

118. ^ *The Code of 1650*, 31 & *The Colonial Laws of Massachusetts*, 55 & 128.

119. ^ Dubois, William E. B. *The Suppression*, Kindle Location 960 & George Moore, *Notes*, 74 & Henry Wilson, *History*, Vol 1, p. 6-7.

120. ^ Russell, John H. *The Free Negro in Virginia*, 38-39.

121. ^ Moore, George. *Notes,* 142 & Henry Wilson. *History,* Vol 1. p. 4.

122. ^ Dubois, William E. B. *The Suppression,* Kindle Location 392.

123. ^ Hinde, Thomas. "Speech of Patrick Henry." 129-132.

124. ^ *Proceedings of the Convention of Delegates,* 4.

125. ^ Levy, Andrew. *First Emancipator,* 75 & 90.

126. ^ Sedgewick, Catherine. "Slavery in New England. " 421 & William Nell, *Colored Patriots,* 52-58.

127. ^ Nell, William. *Colored Patriots,* 59.

128. ^ Hardin, William. *Litigating the Lash,* 160-161 & Fritz Hirschfeld, *George Washington,* 193-194.

129. ^ Armistead, Wilson. *Anthony Benezet,* 78-80.

130. ^ National Archives. "To George Washington From Warner Mifflin: November 23, 1792."

131. ^ U.S. Congress, *Annals of Congress,* 2nd Congress, Session 2, p. 730-731 & Thomas Benton, et al. *Debates of Congress,* Vol 1. p. 397.

132. ^ Franklin, Benjamin. *Works of Benjamin Franklin.* Vol 5, 153-155 & Benjamin Franklin, *Memoirs,* Vol 1. 388-389 & James Madison, *Writings of James Madison: Journal of the Constitutional Convention,* Vol 3. 309-312.

133. ^ Franklin, Ben. *Memoirs,* Vol 1. 388-389.

134. ^ Kaminski, John. *Necessary Evil?* 42-114 & Hirschfeld, *Washington & Slavery,* 173-175 & Livermore, Historical Research, 64-100 & Max Farrand, *Records,* Vol 2, 364-379.

135. ^ Wilson, *History,* Vol 1. p. 52.

136. ^ Kaminski, *Necessary Evil?* 163.

137. ^ Franklin, Ben. *Works,* Vol 5. p. 155-157.

138. ^ Kaminski, *Necessary Evil?* 42-114 & Livermore, *Historical Research,* 64-100.

139. ^ Wilson, *History,* Vol 1. p. 61

140. ^ Hirschfeld, Fritz. *George Washington and Slavery,* 181-183 & John Kaminski, *A Necessary Evil?* 201-229 & Livermore, *Historical Research,* 37-38 & Henry Wilson, *History,* Vol 1, p. 62-67.

141. ^ Hirschfeld, Fritz, *George Washington and Slavery,* 181-183 & John Kaminski, *Necessary Evil?* 201-299 & Livermore, *Historical Research* 37-38 & Wilson, *History,* Vol 1. p. 62-67.

142. ^ Benton and Rives, eds. *Abridgment of the Debates of Congress,* 207-239 & John Kaminski, *Necessary Evil?* 201-230.

143. ^ Dubois, William E. B. *The Suppression,* Chapter 7 & John Kaminski, *Necessary Evil?* 201-230 & Henry Wilson, *History,* Vol 1. p. 9, 62-67.

144. ^ Levy, Andrew. *The First Emancipator*, 119 & 123.

145. ^ Ibid.

146. ^ Levy, Andrew. *The Anti Jefferson*, 23.

147. ^ Levy, Andrew. *The First Emancipator*, 144.

148. ^ Ibid, 169.

149. ^ Levy, Andrew. *The Anti Jefferson*, 25.

150. ^ Levy, Andrew. *The First Emancipator*, 155.

151. ^ Ibid, 163.

152. ^ Ibid, 168.

153. ^ Ibid, 177.

154. ^ Hirschfeld, Fritz. *George Washington and Slavery*, 213 & John Kaminski, *Necessary Evil?* 28 & Henry Wiencek, *Imperfect God*, 272.

155. ^ Jordan and Walsh, *White Cargo*, 41-42, & 79.

156. ^ Vaver, Anthony. *Bound*, Kindle Location 231.

157. ^ Jordan and Walsh, *White Cargo*, 268 & Vauver, *Bound*, Kindle Loc. 268.

158. ^ Vaver, *Bound*, 3213.

159. ^ Ibid, 2341.

160. ^ Jordan and Walsh, *White Cargo*, 280.

161. ^ Sparks, *Life*, 408.

162. ^ U.S. Continental Congress, *Journals*, Vol 3. p. 290.

163. ^ Adams, John and Abigail. *Letters*. September 16, 1774.

164. ^ Coburn, *Battle*, 63.

165. ^ Declaration of Independence.

166. ^ Adams, John and Abigail, *Letters*, June 17, 1776. July 3 & 7, 1776.

167. ^ Lafayette, *Memoirs*, Vol 1. p. 19.

168. ^ Washington, *Writings*, Vol 5. p. 329 & Johnson, *Washington*, 76.

169. ^ Lafayette, *Memoirs*, Vol 1. p. 19.

170. ^ Unger, *Lafayette*, 51.

171. ^ Lafayette, *Memoirs*, Vol 1. 47-48 & Mack and Cloquet, *Life*, 77-78 & Unger, *Lafayette*, 76-77.

172. ^ Ibid.

173. ^ Washington, *Writings*, Vol 1. p. 300.

174. ^ Unger, *Lafayette*, 58.

175. ^ Washington, *Writings*, Vol 5. p. 197-199.

176. ^ Lafayette, *Memoirs*, Vol 2. p. 25.

177. ^ Ibid. p. 61.

178. ^ Unger, *Lafayette*, 189.

179. ^ Malone, Dumas. *Jefferson and the Rights of Man*. Vol 2. p. 40.

180. ^ Ibid, 46.

181. ^ Lafayette, *Memoirs,* Vol 2. p. 55, 70 & Livermore, *Historical Research,* 40-41 & Hirschfeld, *George Washington,* 123-124.

182. ^ Ibid.

183. ^ Unger, *Lafayette,* 193.

184. ^ Livermore, *Historical Research,* 40-41.

185. ^ Lafayette, *Memoirs,* Vol 2. p. 128 & Unger, *Lafayette,* 215-216.

186. ^ Lafayette, *Memoirs,* Vol 2. p. 140.

187. ^ Unger, *Lafayette,* 221. (Note: the quote from Lord John Russell is from Lovejoy, Moore, et. al, *Owen Lovejoy Speeches,* p. 317).

188. ^ Ibid, 222.

189. ^ Jefferson, *Writings,* Vol 3. p. 111-117.

190. ^ Thiers, Adolphe. *The History,* 16.

191. ^ Unger, *Lafayette,* 258.

192. ^ Ibid, 239.

193. ^ Lafayette, *Memoirs,* Vol 2. p. 192.

194. ^ Unger, *Lafayette,* 281.

195. ^ Morris, *Letters,* Vol 1. 566, 571, 582-584.

196. ^ Ibid, 566 & 586.

197. ^ Crawford, *Madam De Lafayette,* 197-201 & Livermore, *Historical Research,* 41.

198. ^ Unger, *Lafayette,* 328.

199. ^ Sumner, Charles. "Lafayette: The Faithful One." 31-32.

200. ^ Unger, *Lafayette,* 355.

201. ^ Levasseur, *Lafayette in America,* Vol 2. p. 37-38.

202. ^ Ibid.

203. ^ Mack and Cloquet. *Lafayette,* 324.

204. ^ Levasseur, *Lafayette,* Vol 2. p. 249-254.

205. ^ Unger, *Lafayette,* 365.

206. ^ Sumner. "Faithful One." 32.

207. ^ Unger, *Lafayette,* 369.

208. ^ Kennedy, Melvin. *Lafayette and Slavery,* 30-32 & 40-42.

209. ^ Ibid.

210. ^ Sumner, Charles. "Lafayette: The Faithful One" 28.

211. ^ Kennedy, *Lafayette and Slavery,* 41-42 & Garrison, *Letters,* 192.

212. Adams, *Writings,* Vol 1, 5-13.

213. ^ Ibid.

214. ^ Adams, *Memoir,* Vol 8. 246.

215. ^ Adams, *Memoir,* Vol 8. 247.

216. ^ Adams, *Memoir,* Vol 5. 139.

217. ^ Adams, John. Abigail Adams, and Charles F. Adams. *Familiar Letters,* 41.

218. ^ Quincy, Josiah. *Life,* 387 & Wilson, *History,* Vol 2. p. 456 & Vol 3 p. 717.

219. ^ Adams, *Memoir,* Vol 4. 492.

220. ^ Adams, *Memoir,* Vol 5. 10.

221. ^ Adams, *Memoir,* Vol 5, 11. Writing in his diary, Adams likened this to the Bible story of Benjamin in Genesis 49:27. "Their masters are privileged with nearly a double share of (Congressional) representation. Benjamin portioned above his brethren has ravened as a wolf. In the morning he has devoured the prey and at night he has divided the spoil."

222. ^ Adams, *Memoir,* Vol 5. 120.

223. ^ Adams, *Memoir,* Vol 9. p. 418 & Wilson, *Rise,* Vol 2. 162.

224. ^ Adams, *Memoir,* Vol 9. 310 & 350.

225. ^ Globe 24th Congress, Session 1, p. 73.

226. ^ Adams, *Memoir,* Vol 9. 273.

227. ^ Adams, *Speech* 1838, 68 & 83.

228. ^ Globe 24th Congress, Session 1, p. 157-158 & Hammond, *Remarks,* 3-18.

229. ^ Adams, *Memoir,* Vol 9. p. 275.

230. ^ Globe 24th Congress, Session 1, p. 474-475.

231. ^ Adams, *Speech* 1836, 5.

232. ^ Globe 25th Congress, Session 1, p. 498.

233. ^ Adams, *Speech* 1836, 15 & Quincy, *Life,* 243-249.

234. ^ Globe 25th Congress, Session 1. p. 498.

235. ^ Globe 24th Congress, Session 1. p. 337.

236. ^ Barnes, *Impulse,* 266.

237. ^ Adams, *Speech 1838,* 62-78 & Quincy, *Life,* 277-281. See the story of Jael in Judges 4:17-24, Esther in the book of Esther and Lazarus in John 11:1-45.

238. ^ Ibid.

239. ^ Falkner, *President,* 154.

240. ^ Brodie, *Thaddeus Stevens,* 107 & Lerner, *Grimke Sisters,* 34 & Levine, *Fall,* Kindle Location 360 & Marshall and Manuel, *Sea to Sea,* 275.

241. ^ Adams, *Letters from Adams to His Constituents,* 50 & Globe 24th Congress, Session 2, Appendix p. 261-265.

242. ^ Adams, *Letters from Adams to His Constituents,* 50.

243. ^ Adams, *Letters from Adams to His Constituents,* 45-65 & Globe 24th Congress, Session 2, Appendix p. 261-267 & Wilson, *Rise,* Vol 1. p. 346-350.

244. ^ Ibid.

245. ^ Adams, *Letters from Adams to His Constituents,* 45-65 & Globe 24th Congress, Session 2, Appendix p. 261-267.

246. ^ Wilson, *Rise*, Vol 1. p. 425-426.

247. ^ Ibid.

248. ^ Globe 24th Congress, Session 2, Appendix p. 261-265.

249. ^ Globe 26th Congress, Session 1, p 150-151.

250. ^ Globe 27th Congress, Session 2, p. 209 & Globe 24th Congress, Session 2, Appendix p. 265.

251. ^ Breaking upon the wheel was a torture used by England to suppress rebellions. During the Revolutionary War, had any of America's Founding Fathers been captured they would have suffered this fate of having all their bones broken as the wheel turned.

252. ^ Adams, *Memoir*, Vol 9. p 58, 349-350, 365, & 418 & Vol 11. p 86.

253. ^ Adams, *Memoir*, Vol 9. p. 349-350, 418 & Vol 11. p. 79-80.

254. ^ Globe 27th Congress, Session 2, p. 158.

255. ^ Giddings, *History*, 158-159.

256. ^ Barnes, *Impulse*, 184-185 & Falkner, *President*, 244, 249 & Marshall and Manuel, *Sounding Trumpet*, 107-108 & Miller, *Arguing*, 408, 426-427 & Wheelan, *Extraordinary*, 190-191, 195.

257. ^ Globe 27th Congress, Session 2, p. 158.

258. ^ Adams, *Speech 1836*, 5 & Barnes, *Impulse*, 185 & Falkner, *President*, 249-250 & Miller, *Arguing*, 427 & Wheelan, *Extraordinary*, 190-191.

259. ^ South Carolina passed this law fearing that free African Americans would distribute David Walker's book and encourage the slaves to revolt. David Walker was a strong Christian who wrote a powerful book to the slaves, telling them the truth about how God's will was freedom. We will study Walker's life in a future book.

260. ^ Globe 27th Congress, Session 2, p. 158-163.

261. ^ Adams, *Memoir*, Vol 11. p. 71 & Globe 27th Congress, Session 2, p. 168-169.

262. ^ Globe 27th Congress, Session 2, p. 168-169.

263. ^ Globe 27th Congress, Session 2, p. 188-190 & Marshall, Thomas. Speeches and Writings, 142-186.

264. ^ Giddings, History, 162-163 & Globe 27th Congress, Session 2, p. 169-170 & Marshall, *Speeches and Writings*, 142-186.

265. ^ Wilson, *History*, Vol 1. p. 425.

266. ^ Giddings, History, 164-167 & Globe 27th Congress, Session 2, p. 170.

267. ^ Globe 27th Congress, Session 2, p. 170-177.

268. ^ Giddings, *History* 167.

269. ^ Globe, 27th Congress, Session 2, p. 170-177.

270. ^ Adams, *Memoir*, Vol 11. p. 85.

271. ^ Giddings, *History* 167.

272. ^ Globe, 27th Congress, Session 2, p. 183.

273. ^ Giddings, *History*, 171-172.

274. ^ Globe 27th Congress, Session 2, p. 188-194.

275. ^ Adams, *Memoir*, Vol 11. p. 86.

276. ^ Barnes, *Impulse*, 124.

277. ^ Giddings, *History*, 119-121 & Wilson, *History*, Vol 1. p. 391-393.

278. ^ Giddings, *History*, 119-121 & Wilson, *History*, Vol 1. p. 391-393.

279. ^ Giddings, *History*, 217-218 & Wilson, *History*, Vol 1. p. 431-432.

280. ^ Wilson, *History*, Vol 1. p. 434-435.

281. ^ Giddings, *History*, 219.

282. ^ Adams, *Memoir*, Vol 9. p. 497.

283. ^ Adams, *Memoir*, Vol 9. p. 298.

284. ^ Globe 27th Congress, Session 2, p. 429.

285. ^ Adams, *Memoir*, Vol 11. p. 407.

286. ^ Lincoln, Abraham. The Emancipation Proclamation.

287. ^ Adams, *Argument Before Supreme Court*, 11-12 & Wilson, *History*, Vol 1. p. 458.

288. ^ Adams, *Memoir*, Vol 10. p. 453-454.

289. ^ Adams, *Argument Before Supreme Court*, 8-9.

290. ^ Wilson, *History*, Vol 1. p. 468.

291. ^ Adams, *Argument Before Supreme Court*, 3-15, 117-121, & 133-134. Also see Lloyd, *Navy*, 42.

292. ^ Falkner, *President*, 236 & Marshall and Manuel, *Sounding Trumpet*, 106.

293. ^ U. S. Supreme Court. Reports of Decisions, Vol 15. p. 156-165.

294. ^ Falkner, President, 302 & Wheelan, Extraordinary, 244-246.

295. ^ Matthew 8:18-20. Globe 29th Congress, Session 1, p. 339-342.

296. ^ Adams, *Memoir*, Vol 12. p. 242-245 & Globe 29th Congress, Session 1, p. 339-342.

297. ^ Congress, *Addresses*, 3.

298. ^ Quincy, *Life*, 426.

299. ^ Matthew 24:44 & Globe 30th Congress, Session 1, p. 386-387.

300. ^ Birney, Catherine. Grimke Sisters, Kindle Loc 3814.

301. ^ Weld et al. *Letters*, Vol 1. p. 44 & Robert Abzug, *Passionate Liberator*, 86

302. ^ Barnes, *Impulse*, 15 & Weld et al. *Letters*, Vol 1. p 52 & Wyatt-Brown, *Lewis Tappan*, 65.

303. ^ Thomas, *Crusade*r, 34-35.

304. ^ Lesick, *Lane Rebels*, 33.

305. ^ Birney, William. *James G. Birney*, 107, 149, 151-152.

306. ^ Beecher, Lyman. *Autobiography,* Vol 2. p 224 & Lesick, *Lane Rebels,* 39 & Marshall and Manuel, *Sounding Trumpet,* 60.

307. ^ Beecher, Lyman. *Autobiography,* Vol 2. p 273-276.

308. ^ Lesick, *Lane Rebels,* 80.

309. ^ Stanton, *Recollections,* 46-47.

310. ^ Bradley, *Brief* & Thome et al. *Debate,* 4.

311. ^ Thome et al. *Debate,* 4.

312. ^ Lyman, Huntingdon. *Lane Seminary Rebels,* 62.

313. ^ Barnes, *Impulse,* 67.

314. ^ Lesick, *Lane Rebels,* 82-83.

315. ^ Ibid, 103.

316. ^ Thome is referring to the story in 2Samuel 12:7 where the Prophet Nathan uses a story to confront King David on his affair with another man's wife. Nathan got King David outraged at the injustice in the scenario then told him that he was the man described.

317. ^ Thome, Debate, 7-11.

318. ^ Lyman, Huntingdon. *Lane Seminary Rebels,* 63 & Grimke, Angelina. *Letters to Catherine,* 83-84 & Thome, *Debate,* 6.

319. ^ Barnes, *Impulse,* 68 & Lesick, *Lane Rebels,* 88-89 & Marshall and Manuel, *Sounding Trumpet,* 72-73.

320. ^ Lerner, *Grimké Sisters,* 157-158, Miller, *Arguing,* 90 & Thomas, *Crusader,* 73-74 & Weld et al. *Letters,* Vol 1. p 273.

321. ^ Thomas, *Crusader,* 74 & Weld et al. *Letters,* Vol 1. p 134.

322. ^ Lesick, *Lane Rebels,* 93.

323. ^ Henry, *Lane Rebels,* 3.

324. ^ Lesick, *Lane Rebels,* 93.

325. ^ Lesick, *Lane Rebels,* 91-92 & Weld et al. *Letters,* Vol 1. p. 137-146.

326. ^ Beecher, Lyman. *Autobiography,* Vol 2. p 325.

327. ^ Barnes, *Impulse,* 71 & Lesick, *Lane Rebels,* 125 & Weld et al. *Letters,* Vol 1. p 173.

328. ^ Mahan, Asa. *Intellectual,* 175-176.

329. ^ Beecher, Lyman. *Autobiography,* Vol 2. p 321.

330. ^ Henry, *Lane Rebels,* 11.

331. ^ Beecher, Lyman. *Autobiography,* Vol 2. p 324 & Thomas, *Crusader,* 72 & Lesick, *Lane Rebels,* 89 & Weld et al. *Letters,* Vol 1. p 133.

332. ^ Thomas, *Crusader,* 78 & Lesick, *Lane Rebels,* 90.

333. ^ Lesick, *Lane Rebels,* 161.

334. ^ Beecher, Lyman. *Autobiography,* Vol 2. p 326.

335. ^ Lesick, *Lane Rebels*, 92.

336. ^ Mahan, Asa. *Intellectual,* 172-175.

337. ^ Ibid, 187.

338. ^ Ibid, 186-187.

339. ^ Ibid, 179.

340. ^ Beecher, Lyman. Autobiography, Vol 2. p 327-328.

341. ^ Ibid.

342. ^ Mahan, Asa. *Intellectual,* 186.

343. ^ Lyman, Huntingdon, *Lane Seminary Rebels,* 64

344. ^ Lesick, *Lane Rebels,* 129.

345. ^ Weld, *Lane Rebels Statement,* 20-26.

346. ^ Ibid, 26-27.

347. ^ Ibid, 16-17.

348. ^ Beecher, Lyman. *Autobiography,* Vol 2. p 321.

349. ^ Weld et al. *Letters,* Vol 1. p 152.

350. ^ Lesick, *Lane Rebels,* 138 & Weld et al. *Letters,* Vol 1. p 183.

351. ^ Lesick, *Lane Rebels,* 133 & Weld et al. *Letters,* Vol 1. p 341.

352. ^ Bradley, *Brief.*

353. ^ Birney, William. *James G. Birney,* 136.

354. ^ Barnes, *Impulse,* 235.

355. ^ Barnes, *Impulse,* 237 & Dumond, *New Views,* 379-380.

356. ^ Beecher, Lyman. *Autobiography,* Vol 2. p 321.

357. ^ Barnes, *Impulse,* 86 & Falkner, *President,* 145 & Marshall and Manuel, *Sounding Trumpet,* 79 & Weld et al. *Letters,* Vol 1. p XXIII.

358. ^ Lundy and Earle, *Life,* 280.

359. ^ Thomas, *Crusader,* 104 & Weld et al. *Letters,* Vol 1. p 206-207.

360. ^ Barnes, *Impulse* 80-81.

361. ^ Barnes, *Impulse,* 81 & Marshall and Manuel, *Sounding Trumpet,* 79 & *Thomas, Crusader,* 103.

362. ^ Beecher, Lyman. *Autobiography,* Vol 2. p 345.

363. ^ Goodell, William. *Slavery,* 426-427.

364. ^ Lundy and Earle, *Life,* 279-280.

365. ^ Weld et al. *Letters,* Vol 1. 154-155.

366. ^ Thomas, *Crusader,* 23.

367. ^ Dumond, *New Views,* 383-384 & Lesick, *Lane Rebels,* 179 & Marshall and Manuel, *Sounding Trumpet,* 83 & Thomas, *Crusader,* 109 & Weld et al. *Letters,* Vol 1. p 318-319.

368. ^ Lesick, *Lane Rebels,* 179 & Weld et al. *Letters,* Vol 1. p 318-319.

369. ^ Lesick, *Lane Rebels*, 179-180 & Weld et al. *Letters*, Vol 1. p 326-329.

370. ^ Lesick, *Lane Rebels*, 177 & Weld et al. *Letters*, Vol 1. p 242-245.

371. ^ Weld et al. *Letters*, Vol 1. p. 100.

372. ^ Dumond, *New Views*, 381 & Grimstead, *American Mobbing*, 33, 56-57, & Thomas, *Crusader*, 105-106 & Weld et al. *Letters*, Vol 1. p 260-262.

373. ^ Stanton, *Recollections*, 49-55.

374. ^ Wilson, *History*, Vol 1. p 367-368.

375. ^ Lesick, *Lane Rebels*, 173.

376. ^ Dresser, *Narrative*, 6-14.

377. ^ Thomas, *Crusader*, 104-105.

378. ^ Thomas, *Crusader*, 93 & Weld et al. *Letters*, Vol 1. p 182.

379. ^ Weld et al. *Letters*, Vol 1. p 182.

380. ^ Lesick, *Lane Rebels*, 182.

381. ^ Dumond, *Antislavery*, 163-164.

382. ^ Coffin, Levi. *Reminiscences*, 223-230.

383. ^ Weld, *Bible Against Slavery*, 10-12.

384. ^ Weld, *Bible Against Slavery*, 68-71.

385. ^ Thomas, *Crusader*, 131.

386. ^ Globe 24th Congress, Session 1, p. 157-171 & Hammond, *Remarks*, 1-20.

387. ^ Wilson, *History*, Vol 1. p 424.

388. ^ Davis, Varina. *Jefferson Davis*, Vol 1. 419-426, 432-436.

389. ^ Ibid.

390. ^ Birney, William. *James G. Birney*, 190.

391. ^ Globe 24th Congress, Session 1, p. 221.

392. ^ Ibid.

393. ^ Keitt, *Speech*, 1-22. Those statistics are from reports released by the Secretary of the Treasury.

394. ^ Wilson, *History*, Vol 1. p. 392.

395. ^ Birney, William. *James G. Birney*, 191 & Wilson, *History*, Vol 1. p. 339-341.

396. ^ Ibid.

397. ^ Brisbane, *Slaveholding Examined in Light of the Bible*, iii.

398. ^ Ibid, iv.

399. ^ Lesick, *Lane Rebels*, 194.

400. ^ Lesick, *Lane Rebels*, 186-187 & Weld et al. *Letters*, Vol 1. p. 341

401. ^ Barnes, *Impulse*, 183 & Falkner, President, 244, Marshall and Manuel, *Sounding Trumpet*, 107 & Miller, *Arguing*, 407.

402. ^ Barnes, *Impulse*, 187.

403. ^ Barnes, *Impulse*, 123.

404. ^ Barnes, *Impulse*, 186 & Falkner, *President*, 259 & Thomas, *Crusader*, 206 & Weld et al. *Letters*, Vol 2. p. 905-906.

405. ^ Rushing, Kittrell. *Memory and Myth*, 8.

406. ^ Coffin, *Reminiscences*, 610-618.

407. ^ Ibid, 607-609.

408. ^ Thomas, *Crusader*, 252.

409. ^ Lyman, Huntingdon. *Lane Seminary Rebels*, 68-69.

410. ^ Barnes, *Impulse*, XXV.

About The Author

Rebekah Brewster is a professional writer who loves making words come alive on the page. She is a California girl with a Midwestern heart. Her hobbies are reading lots and lots of books, cooking, watching UFC fights and hiking. She loves this country and all those who sacrifice for it.

She cares about her readers and hopes this book inspires you to be strong during difficult times.

If you enjoyed the book, please leave a good review on the website where purchased.

Rebekah is also the author of Walk With God: 25 True Stories and Troublemakers.

If you would like Rebekah to speak at your church, you can email her at info@quietbeauty.org.